JOURNEY INTO THE WHIRLWIND

Eugenia Semyonovna Ginzburg · Journey into the Whirlwind ·

Translated by PAUL STEVENSON *and* MAX HAYWARD

A Harvest Book
A Helen and Kurt Wolff Book
Harcourt, Inc.
San Diego   New York   London

For information about permission to reproduce
selections from this book, please write Permissions,
Houghton Mifflin Harcourt Publishing Company
215 Park Avenue South, NY, NY 10003.

Originally published in Italy: in Russian, as *Krutoj Mars rut;*
in Italian, as *Viaggio nella Vertigine.*

Library of Congress Cataloging-in-Publication Data
Ginzburg, Evgeniia Semenovna.
Journey into the whirlwind.
"A Helen and Kurt Wolff book."
Translation of Kruto marshrut.
1. Political prisoners—Russia—Personal
narratives. 2. Russia—Social conditions—1917—
I. Title.
[DK268.3.G513 1975]   365¢.6¢0924[B]   74-16406
ISBN 0-15-602751-8 (Harvest: pbk.)

Printed in the United States of America
First Harvest edition 1975

DOC 20 19 18 17 16 15 14 13

CONTENTS

PART TWO

And so I appeal to our government:
double it, treble it, the guard on this grave.

—*Yevtushenko*

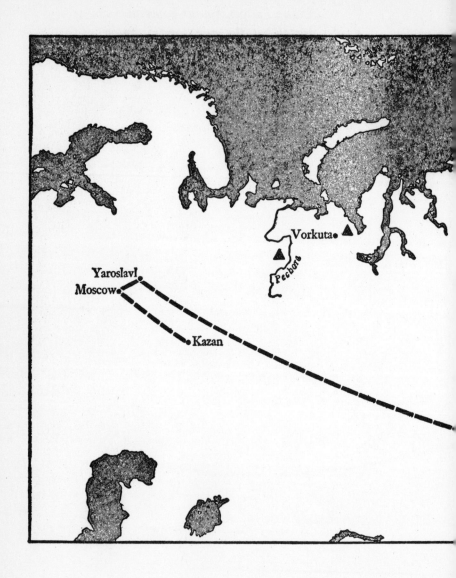

Yaroslavl

Moscow

Vorkuta

Pechora

Kazan

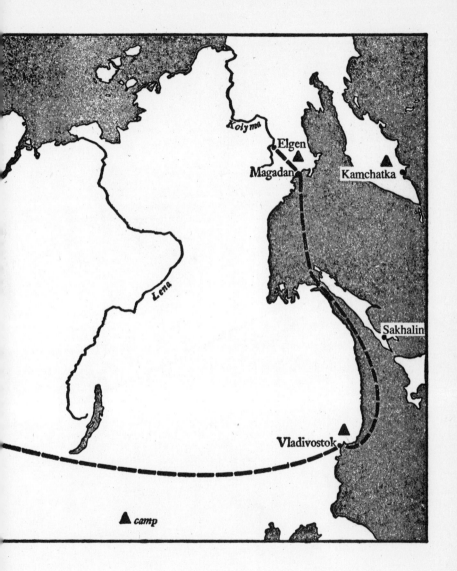

Kolyma

Elgen

Magadan

Kamchatka

Lena

Sakhalin

Vladivostok

▲ camp

Part one

1

• *A telephone call at dawn*

The year 1937 began, to all intents and purposes, at the end of 1934—to be exact, on the first of December.

At four in the morning the telephone bell rang shrilly. My husband, Pavel Vasilyevich Aksyonov, a leading member of the Tartar Province Committee of the Party, was away on business. I could hear my children in the next room breathing evenly in their sleep.

"You're wanted at the regional committee office, Room 37, at six A.M."

This order was given me as a member of the Party.

"Is it war?"

But they had hung up. Clearly, in any case, there was some sort of serious trouble.

Without waking anyone, I ran out of the house long before there was any traffic in the street. I can still remember the silent snowfall and the strange lightness of my walk.

I don't want to sound pretentious, but I must say in all honesty that, had I been ordered to die for the Party—not once but three times—that very night, in that snowy winter dawn, I would have obeyed without the slightest hesitation. I had not the shadow of a doubt of the rightness of the Party line. Only Stalin—I suppose instinctively—I could not bring myself to idolize, as it was already becoming the fashion to do. But if I felt this vague disquiet about him, I carefully concealed it even from myself.

There were already about forty fellow teachers, Com-

munists—colleagues of mine and people I knew—crowding
in the corridors outside the office. Pale and silent, they
had all been wakened like me in the middle of the night.
We were waiting for the regional committee secretary,
Lepa.

"What has happened?"

"Don't you know? Kirov * has been murdered."

Lepa, a stolid, usually imperturbable and inscrutable
Latvian, a Party member since 1913, was clearly not himself. He spoke for less than five minutes. He knew nothing
whatever of the circumstances of the murder and merely
repeated the words of the official communiqué. He had
summoned us in order to send us out to various factories,
to address the workers and give them a brief account of
the situation.

I was assigned to a textile mill at Zarechye, the industrial
district of Kazan. Standing on a pile of cotton-filled sacks
in the middle of the factory floor, I conscientiously repeated what Lepa had told us to say, but my thoughts were
in such turmoil that I could hardly keep my mind on the
message.

When I got back to town, I dropped in at the committee
building for a glass of tea in the canteen. Sitting next to
me was Yestafyev, the director of the Marxist Institute.
He was a good, simple man of proletarian origin from
Rostov who had been a member of the Party since before
the Revolution. In spite of the twenty years' difference in
our ages, we were friends and had interesting talks whenever we met. Now he drank his tea in silence, without
looking at me. Then he glanced over his shoulder, leaned

* Sergey Mironovich Kirov, b. 1866, Secretary of the Central Committee of the Communist Party of the Soviet Union, whose assassination
in Leningrad on December 1, 1934, was used as the excuse for the great
purges in the following years.

toward me, and said in a voice so strange, so unlike his own, that it filled me with a terrible foreboding of misfortune:

"The murderer—he was a Communist, you know."

2

• *The red-haired professor*

The long indictments in the case of Kirov's murder, published in the newspapers, made one's blood run cold but did not at the time give rise to misgivings. It was fantastic, unheard of, that the guilty men—Nikolayev, Rumyantsev, Katalynov—should be former members of the Leningrad Komsomol, but *Pravda* said so, and it must therefore be true.

Then the repercussions spread like the ripples when a stone is thrown into a quiet pool.

On a sunny day in February 1935, Professor Elvov came to see me. He had appeared on the Kazan University scene after an unpleasantness connected with the four-volume *History of the All-Union Communist Party (Bolshevik)*, edited by Yemelyan Yaroslavsky.* The chapter on the events of 1905, contributed by Elvov, was found to contain certain errors in its treatment of the theory of permanent revolution. The whole book and Elvov's article in

* Eminent Party theoretician, 1878–1943.

particular had been condemned by Stalin in his famous letter to the editor of the review *Proletarian Revolution*. After the appearance of the letter, the errors were defined more specifically as smuggled-in "Trotskyist ideas."

But in those days, before the shooting of Kirov, matters of this sort were taken much less tragically. Elvov, with the approval of the Party, proceeded to Kazan, where he took up a professorship at the Teachers Training Institute. He was elected to the Party's municipal committee and became a speaker at meetings of Party activists and of the city's intelligentsia. Indeed, it was he who addressed the activists' group on the occasion of Kirov's murder.

Elvov was a striking-looking man with a shock of curly red hair and a large head sitting squarely on his shoulders. He had almost no neck, so that his big, sturdy body gave an impression both of strength and of a sort of helplessness. Wherever he went his appearance and intellectual qualities attracted attention.

His brilliant and at times grandiloquent lectures, his trenchant and dogmatic speeches, the floods of erudition he poured out upon the timid heads of his local colleagues, had made him one of the most hated men in town. In 1935 he was thirty-three.

There he sat opposite me on that frosty, sunny February day in 1935—not in the armchair beside the desk but on a plain chair in the corner of the room; his legs in their elegant boots were not sprawling as usual but tucked up under the chair. His complexion, too, was not its usual pink and white but a dirty gray. Holding on his lap my two-year-old son Vaska, who had run into the room, he was saying through bluish, trembling lips:

"I've got one too, you know. Sergey. He's four . . . a good little fellow. . . ."

Later, I was often to see eyes with the same expression as Elvov's that day. There was an indefinable mixture in them of pain, anxiety, the weariness of a hunted animal, and somewhere deep down a half-crazy glint of hope. I too must have looked like that in the years that followed, but I cannot be sure, for the simple reason that in all those years I never once saw my face in a looking glass.

"What's wrong, Nikolay Naumovich?"

"I'm in for it, that's all. . . . I just dropped in for a minute to tell you, so that you shouldn't think . . . It's all a pack of lies. I swear I never did anything against the Party."

I remember now with shame the bromides with which I tried to comfort him. Surely things couldn't be as bad as all that. At the worst, in view of the general situation at the moment, he might get a delayed reprimand for that unfortunate article . . . and so on. . . .

When I had finished he said something quite startling:

"I can't tell you how sorry I am that you may suffer because of your association with me. . . . I wouldn't have had it happen for the world."

I looked at him in blank amazement. Was he out of his mind? Suffer for my association with him! What association was he talking about? What nonsense!

We had first met immediately after his arrival in Kazan, I believe in autumn 1932. At the time I was working at the Teachers Training Institute, where he became assistant professor of Russian history and was given an apartment in the Institute building. He immediately made plans for several joint publications, and various scholars would meet at his apartment to discuss them. I was invited, I remember, to take part in preparing a source book on the history of Tartary.

Later I worked with Elvov again, at the editorial office of the local paper, *Red Tartary*. After a row between the new editor, Krasny, and the old members of the staff, the regional Party committee decided to bring new blood in by recruiting a few intellectuals. I was made assistant head of the cultural department, and Elvov of the department of foreign news.

Since when, I asked him, had the fact of working together at a Soviet university and on a Party newspaper become an "association," let alone something one could "suffer" for?

Evidently, at this dreadful moment in his life, stripped of all complacency and pose, he had been endowed with the gift of understanding, for what he rightly saw in my words was not cowardice or hypocrisy but an unshakable political naïveté. Yes, I was a Party member, a historian and a writer, I was safely launched on an academic career; politically, I was a babe in arms, and this he realized.

"You don't understand what's going on. That will make things hard for you—even harder than for me. Good-by."

He stood in the hall, struggling awkwardly into the sleeves of his leather coat. My elder son, Alyosha, then nine years old, watched him gravely from the doorway, then came up and helped him. When the door had closed behind the "red-haired professor" Alyosha said:

"You know, Mother, I don't really like him very much. But I think he's in very bad trouble now. I'm sorry for him."

When I arrived at the Institute to lecture next morning, the old porter who had known me since my student days rushed up to me in the hall:

"You know our professor, the red-haired one . . . They took him away in the night. Under arrest."

3

The next two years might be called the prelude to that symphony of madness and terror which began for me in February 1937. A few days after Elvov's arrest, a Party meeting was held at the editorial office of *Red Tartary* at which, for the first time, I was accused of what I had *not* done.

I had *not* denounced Elvov as a purveyor of Trotskyist contraband. I had *not* written a crushing review of the source book on Tartar history he had edited—I had even contributed to it (not that my article, dealing with the nineteenth century, was in any way criticized). I had *not*, even once, attacked him at a public meeting.

My attempts to appeal to common sense were summarily dismissed.

"But I wasn't the only one—no one in the regional committee attacked him!"

"Never you mind, each will answer for himself. At the moment it's you we are talking about."

"But he was trusted by the regional committee. Communists elected him to the municipal board."

"You should have pointed out that this was wrong. What were you given a university training for, and an academic job?"

"But has it even been proved that he's Trotskyist?"

This naïve question provoked an explosion of righteous anger:

"Don't you know he's been arrested? Can you imagine

anyone's being arrested unless there's something definite against him?"

All my life I shall remember every detail of that meeting, so notable for me because, for the first time, I came up against that reversal of logic and common sense which never ceased to amaze me in the more than twenty years that followed right up to the Twentieth Party Congress, or at any rate the plenum of September 1953.

During a recess I went off to the editorial office. I wanted a moment to myself to think of what I should do next and how to behave without losing my dignity as a Communist and a human being. My cheeks were burning, and for several minutes I felt as if I should go mad with the pain of being unjustly accused.

The door creaked, and Alexandra Alexandrovna, the office typist, came in. She had done a lot of work for me and we got on well. An elderly, reserved woman who had suffered some kind of disappointment in life, she was devoted to me.

"You're taking this the wrong way, Eugenia Semyonovna. You should admit you're guilty and say you are sorry."

"But I'm not guilty of anything. Why should I lie at a Party meeting?"

"You'll get a reprimand anyway. A political reprimand is a very bad thing. And by not saying you repent you make it worse."

"I won't be a hypocrite. If they do reprimand me, I'll fight till they withdraw it."

She looked at me with her kindly eyes surrounded by a network of wrinkles, and repeated the very words Elvov had said to me at our last meeting:

"You don't understand what's going on. You're heading for a lot of trouble."

Doubtless, if the same thing happened to me today, I would "recant." I almost certainly would, for I too have changed. I am no longer the proud, incorruptible, inflexible being I was then. But in those days this is what I was: proud, incorruptible, inflexible, and no power on earth could have made me join in the orgy of breast-beating and self-criticism that was just beginning.

Large and crowded lecture halls were turned into public confessionals. Although absolution was not at all easy to come by—expressions of contrition were more often than not rejected as "inadequate"—the torrent of confessions grew from day to day. Every meeting had its chosen theme. People "repented" for misunderstanding the theory of permanent revolution and abstaining from the vote on the program of the opposition in 1923; for failing to purge themselves of great-power chauvinism; for underrating the importance of the second Five-Year Plan; for having known personally some "sinner" or for liking Meyerhold's theater . . .

Beating their breasts, the "guilty" would lament that they had "shown political short-sightedness" and "lack of vigilance," "compromised with dubious elements," "added grist" to this or that mill, and were tainted with "rotten liberalism."

Many such phrases echoed under the vaulted roofs of public buildings. The press, too, was flooded with contrite articles by Party theorists, frightened out of their wits like rabbits and not attempting to conceal their fear. The power and importance of the NKVD * grew with every day.

---

* Initials of the Russian for People's Commissariat of Internal Affairs (later MVD—Ministry of Internal Affairs) which was responsible for the secret political police, internal security, and the vast network of prisons and concentration camps of the kind described in this book. At the head of the NKVD soon after Mrs. Ginzburg's arrest was Nikolay Yezhov, who was replaced in 1938 by Lavrenti Beria. These

The Party meeting duly reprimanded me for "slacken-ing of political vigilance." Most tenacious in pushing this decision was Kogan, who had replaced Krasny as editor of *Red Tartary* and who made a prosecutor's speech arraign-ing me as "potentially sharing Elvov's ideas"!

Soon afterward, it turned out that Kogan himself had once been an oppositionist, and that his wife had been Smilga's * private secretary and had taken part in the spec-tacular "send-off" Smilga was given when he went into exile. Hence Kogan's frightening zeal in "unmasking" other Communists, including such political innocents as myself. At the end of 1936 Kogan, by then transferred to Yaroslavl, found the daily expectation of arrest more than he could bear and threw himself under a train.

I was cheered by the fact that the regional committee secretary showed as little "political understanding" as my-self. When my appeal reached the committee, he said:

"What on earth are they reprimanding her for? Every-body knew Elvov. He was trusted by both the regional and the town committees. Is it for walking down the street with him?"

The reprimand was canceled and replaced (on the in-sistence of other members of the board who understood what was required of them "at this stage" better than did the secretary) by a mild reproof for "insufficient vigi-lance."

names are associated with the worst of the mass persecutions between 1937 and Stalin's death.
* Ivan Smilga, b. 1892, Bolshevik leader expelled from the Party as a Trotskyist in 1927. The date of his death is uncertain.

4

• *The snowball*

About four and a half miles out of town, on the picturesque river Kazanka, stood the regional committee's country villa "Livadia," built by Lepa's predecessor, former regional committee secretary Mikhail Razumov. A plump little man with short legs, piercing blue eyes, and a Louis XVI Bourbon profile, he had been a Party member since 1912 and was a close friend of my husband. So we saw a great deal of this "foremost worker in Tartary," as, after the toadying fashion of the day, he was called.

He was a man full of contradictions. An excellent organizer and of unimpeachable loyalty to the Party, he was nevertheless inclined to the cult of his own personality. We had first met in 1929, since when his rise had been meteoric. In 1930 he still lived in one room in my father-in-law's apartment and, when hungry, sliced a sausage with a penknife on a piece of newspaper. By 1930 he had built Livadia, with a cottage for himself on the grounds. In 1933, when the Order of Lenin was conferred on Tartary for its progress in collectivizing the land, his portraits were carried triumphantly through the town, and enterprising artists copied them in any medium from oats to lentils for exhibition at agricultural shows.

As his close friends, we teased him about it long before a similar situation was described by Ilf * and Petrov:

* Ilya Ilf, famous Soviet satirist who, together with Evgeny Petrov, wrote *The Twelve Chairs*.

"The sparrows pecked out your eyes last night, Mikhail Osipovich. Have a look at your portrait at Black Lake."

Members of the regional committee bureau with their families spent their summer holidays at Livadia and went there all year round on their days off.

Once, in the spring of 1935, when we were there, I noticed a new face at one of the tables.

"Who's the red-haired Motele," * I whispered to my husband.

"He's not red but black, and he is Comrade Beylin, the new chairman of the bureau of Party political control."

Little did I think that this cheerful, small-town tailor's face was that of the first of my inquisitors.

We were introduced. I noticed a gleam in his eyes at the mention of my name, but he extinguished it at once by looking down at a dishful of the famous Livadia pastries. I learned later that by then my file was already on his desk.

A few days after our meeting, I sat in Comrade Beylin's office, under his burning, fanatical, and sadistic eyes, while he exercised his Talmudic subtlety in polishing up the definition of my "crimes." The snowball was rolling downhill, growing disastrously and threatening to smother me.

Comrade Beylin spoke quietly, addressing me in the second person singular as was proper between Party members.

"Haven't you read Comrade Stalin's article? With your high qualifications you could hardly have misunderstood it. . . .

"Didn't you know that Elvov had gone astray on the subject of permanent revolution? . . .

"You did not admit your guilt at the Party meeting. Does this mean that you refuse to disarm?"

I was puzzled by this expression and assured him that

* A common Jewish first name.

I had never taken up arms against the Party. His shining
eyes half hooded by soft thick lids, he began again at the
beginning.

"Objectively, anyone who refuses to disarm, when called
upon by the Party to do so, gravitates toward the position
of its enemies. . . ."

I struggled desperately to keep myself from "gravitating" and reminded my exacting confessor that I had not,
after all, done anything wrong—I had merely known Elvov
as a colleague, as had everyone else at the university.

"You still refuse to understand that tolerance toward
anti-Party elements leads objectively to disloyalty. . . ."

Ignoring my replies, he kept rolling the snowball on its
way, in accordance with a definite, deliberate plan which
to me was still a mystery.

Before long our daily talks ceased to be held in private:
we were joined by a colleague of Beylin's from Moscow,
whose name I never learned but whom I mentally called
Malyuta, so much did he remind me of the henchman of
Ivan the Terrible. His methods were the opposite of
Beylin's, but as a sadist and casuist he might have been his
double.

Beylin's eyes, under their bulging lids, shone with a
subdued, sardonic joy at the expense of his fellow creatures, while Malyuta's blazed triumphantly. Beylin spoke
softly, in a low-pitched voice, while Malyuta shouted and
swore. True, his oaths were far from being as fierce as
those I was later to hear from the NKVD. Malyuta's were
political: "Appeasers! Left-right mongrels! Trotskyist
abortions! Mangy opportunists! . . ."

This ordeal went on for two months, by which time
I was on the verge of a nervous breakdown, made worse
by an attack of malaria.

When I compare my experience during this "prelude"

with what I endured later, from 1937 until Stalin's death, or rather till the plenum of July 1953 at which Beria was denounced, I am always struck by the incongruity between my reactions and their outward cause. After all, until the fifteenth of February 1937, my sufferings were only moral. The outward circumstances of my life were unchanged, my family was still safe, and my beloved children were still with me. I lived in my familiar apartment, slept in a clean bed, had plenty to eat and work of an intellectual kind. Yet subjectively I suffered more than in later years, when I was kept in solitary confinement in a security prison, or felled trees in the forests of Kolyma.

Why was this? Perhaps because waiting for an inevitable disaster is worse than the disaster itself, or because physical pain dulls mental anguish. Or perhaps simply because human beings can get used to anything, even to the most appalling evils, so that the successive wounds inflicted on me by the dreadful system of baiting, inquisition, and torture hurt me less than those I suffered when I first came up against it. Be that as it may, 1935 was a frightful year for me. My nerves were at breaking point, and I had persistent thoughts of suicide.

I was cured—temporarily at least—in the early autumn of that year by the tragic story of the Communist Pitkovskaya. She worked in the schools department of the regional committee, and was one of those who had carried over into the thirties all the habits and attitudes of the period of the civil war—one of those of whom Pilnyak * had said: "Communists . . . in leather jackets . . . who go at it all out . . ."

I cannot now recall her name and patronymic—no one

---

* Boris Pilnyak, b. 1894, leading Soviet writer, from whose novel *The Naked Year* this quotation comes. He disappeared in the purges of 1937.

called her by anything but her surname, Pitkovskaya. You
could load her with enough work for four people, you
could take her money and not return it, you could laugh
at her—she never took offense at those whom she called
her family. If anyone thought of the whole Party as one
great brotherhood, it was Pitkovskaya. Selfless by nature,
she burdened her sensitive conscience with a permanent
feeling of guilt as regards the Party. The reason for it was
that in 1927 her husband Dontsov had, for a time, joined
the opposition. She loved him dearly, tenderly, yet she
ruthlessly condemned his past. Even to her five-year-old
son she tried to explain in simple words how gravely his
father had sinned. She insisted that Dontsov should "steep
himself in the proletarian spirit," which in practice meant
that she would not let him live in a big town like Kazan
but made him work at a lathe in the shipyard at Zelenodolsk.

In the autumn of 1935 the authorities began to arrest
everyone who had ever been connected with the opposition. Hardly anyone at the time realized that purges of
this sort were carried out strictly in accordance with a
prearranged plan which affected this or that category of
people quite irrespectively of the way they had actually
behaved. Least of all could Pitkovskaya have understood
such a thing.

When the NKVD came in the middle of the night to
arrest her husband, who had spent Sunday with her in
Kazan, she carried on in a manner worthy of a Greek
tragedy. Needless to say, she was heartbroken for her
beloved husband, the father of her child, but she suppressed
her feelings.

"So he lied to me," she exclaimed dramatically. "So he
really was against the Party all the time!"

With an amused grin, the men from the NKVD said: "Better get his things together."

But she refused to do this for an enemy of the Party, and when her husband went to his sleeping child's cot to kiss him good-by, she barred his way:

"My child has no father!"

Then, shaking the policemen fervently by the hand, she swore to them that her son would be brought up a loyal servant of the Party.

All this she told me herself, and I do not for a moment believe that there was the least calculation or hypocrisy in her actions. Absurd as they seemed, they were prompted by what she genuinely felt in her naïve soul, utterly devoted to the ideals of her militant youth. The idea of possible degeneration, of scoundrels lusting for power, of treachery, of Bonapartism, had no place in her honest, single-track mind.

The day after her husband's arrest she lost her job with the regional committee. She had no specialized training, and she would in any case have found it very hard to get work after being dismissed "for association with an enemy of the Party." For this crime, she herself was soon expelled from the Party.

I lent her my overcoat and some money for her journey to Moscow, where she went in order to get herself reinstated. She did not succeed.

For a short time after her return to Kazan, she worked at a typewriter factory; then she injured her right hand.

After that she was penniless. Her son was expelled from his kindergarten, and by degrees people began to cut her in the street. I learned to recognize her timid, uncertain ring at the doorbell. We fed her and tried to comfort her. Then my husband pointed out that I myself was under suspicion and that association with Pitkovskaya might well prejudice my case. I went through torments of conscience. My natural desire to help a good friend and a devoted Communist fought with the craven fear that if Beylin and his confederate got to know of her daily visits they would tear me to pieces.

Then suddenly she stopped coming to see us. For two or three days we had no news and on the fourth we heard that, after writing a letter full of love and devotion to Stalin, Pitkovskaya had drunk a glass of acetic acid. In her suicide note she blamed nobody, treated the whole thing as a misunderstanding, and begged to be remembered as a Communist.

Her coffin was followed by her five-year-old son, the cleaning woman from the regional committee office, and two or three of the more "desperate characters" among her old friends.

As I looked at the sad little grave, surmounted by neither cross nor star, I told myself: No, I won't do that. I shall put up a fight. They may kill me if they can—but I won't help them.

By the autumn Beylin and Malyuta had announced their verdict: a severe reprimand and warning for having "compromised with hostile elements," and the withdrawal of my license to teach.

But this of course was not the end. The snowball continued to roll downhill.

• *There's no one so silly as a clever man*

My mother-in-law, Avdotya Vasilyevna Aksyonova, a simple, illiterate peasant woman born in the days of serfdom, was of a deeply philosophical cast of mind and had a remarkable power of hitting the nail on the head when she talked about the problems of life. She spoke with the singsong accent of the south and was always coming out with quaint proverbs and sayings. Just as, we are told, King Solomon observed at moments of crisis, "This too will pass," so Grandmother, on being told of some extraordinary event, would usually say: "Yes, it's happened before. . . ."

I remember how startled we were when, as we all sat at table, she commented on Kirov's murder:

"It's happened before, you know."

"What do you mean?"

"Well, it has. They killed the Tsar." (She was thinking, if you please, of the murder of Alexander II.) "Of course I was still a little girl then. Only this time it looks as if they've shot the wrong man. It's Stalin who's the Tsar now, not Kirov—so why shoot Kirov? Oh well, we'll see what happens. . . ."

I remember in the utmost detail the first of September 1935 when, having been forbidden to teach, I shut myself up in my room and went through the torments of Tantalus. All my life I had been either a student or a teacher, and the first day of the academic year had always been for me more important even than the New Year. And here

I sat alone, rejected, while from the street rose the familiar sounds of schools and universities coming back to life again after the summer. Kazan, a students' town, was buzzing with activity. But I would never again walk through the pillared entrance of my own university.

Grandmother was shuffling noisily and sighing outside the door. But I could not bring myself to go out or call her. At the moment I couldn't bear to see anyone, not even the children. I felt as much on my own as Robinson Crusoe.

So I sat until dinnertime, when there came a loud ring at the door and I heard Grandmother's voice saying urgently:

"It's for you, Genia. Come here a minute. . . ."

In the doorway stood a messenger, a boy I had never seen before, and he handed me a large bunch of sad autumn flowers, asters. In it was an affectionate note from my last year's students.

It was more than I could bear, and even before the boy had gone I burst into loud sobs, howling and wailing like a peasant woman, so that Grandmother joined in the chorus, saying again and again: "Poor lamb! My darling!"

Then she suddenly broke off, shut the door, and whispered:

"They're a plucky lot, the students. They'll catch it for those flowers. Genia, darling, I'll tell you something. . . . And you listen to me even though I'm old and ignorant. They're setting a trap for you, Genia, and you'd better run while you still can, before they break your neck. 'Out of sight, out of mind,' they say. The farther away you are the better. Why not go to our old village, to Pokrovskoye?"

I was still sobbing and could hardly follow what she was saying.

"You know, in a place like that they need educated people like you. Our cottage is standing there empty, boarded up. And there are apple trees in the garden. . . . Fifteen of them."

"But how can I, Grandmother? How can I leave everything, the children, my work?"

"Well, they've taken your job away anyhow. And the children won't come to any harm with us."

"But I must prove my innocence to the Party. How can I, a Communist, hide from the Party?"

"Genia, darling, don't talk so loud. And don't be cross with me. I'm not a stranger. Who are you going to prove your innocence to? They used to say, God and the Tsar are far away. So they still are, God and Stalin. . . ."

"No, Grandmother, don't. . . . Even if it kills me, I'll prove it! I'll go to Moscow. I'll fight. . . ."

"Oh, Genia, Genia darling! They say there's no one so silly as a clever man."

My husband smiled condescendingly when I told him of his mother's suggestion. His reaction was quite natural—didn't we possess the truth in its final form, while she was only a simple peasant woman?

Later, when I went to Moscow to plead my cause with the Party control commission, I was offered another suggestion much like my mother-in-law's.

There, at the Party control building in Ilyinka Street, many Communists, first victims of the witch hunt, were to be met in those days. In a queue outside the office of the Party investigator I came across a young doctor, Dikovitsky, whom I had known since my early youth; he was of Gipsy origin. He told me in confidence about his own predicament. He too had shown signs of "lack of vigilance," of "rotten liberalism," of "objectively gravitating," and the rest of it.

"Listen, Genia," he said to me, "we're both in for it and we're not likely to do ourselves any good here. We must think of something else. What would you say, for instance, to going off and living with the raggle-taggle Gipsies, O?"

His eyes with their bluish whites flashed as impudently as in the past.

"You can still joke!"

"But I'm dead serious. Listen. I'm a Gipsy by birth, and you could easily pass for one. Why don't we just drop out of circulation for a bit? Not a word to anyone, not even to our families. Or there could be a black-edged notice in the newspaper, saying P. V. Aksyonov announces with sorrow the untimely death of his beloved companion, and so on. I bet your Beylin would have to close his file, whether he liked it or not. And you and I would join a Gipsy camp and wander about for a year or two, like tourists, until all this blows over. What do you say?"

This suggestion too, which was perfectly sensible, seemed to me crazy, deserving only a smile. Yet looking back on it several years later I realized with amazement that many people had saved themselves in just this way. Some disappeared to distant and exotic regions such as Kazakhstan or the Soviet Far East. One of them was Pavel Kuznetsov, a former editor of a Kazan newspaper, who was charged together with me but never arrested because he went off to Kazakhstan, where they failed to find him and eventually stopped looking. Later, he even published in *Pravda* his translations of odes by Kazakh *akyns* * to "the great Stalin" and "Father Yezhov"!

Some people "lost" their Party cards and, having been

---

* *Akyn:* Central Asian folk poet who composes oral verse.

expelled for this, migrated to other towns or villages. Some women hurriedly became pregnant, naïvely supposing that this would save them from the avenging arm of justice as represented by Yezhov and Beria. These poor wretches gravely miscalculated and only added to the number of deserted orphans.

Yes, people looked for every possible way out. And those in whom common sense, shrewdness, and independence of mind outweighed the effects of a demagogic education and the mystic spell of Party slogans, did in fact sometimes escape.

As for me, I must honestly confess that my way of defending myself—by fervent protestations of innocence and loyalty, vainly made to sadists, or officials who were themselves bewildered by the fantastic course of events and terrified for their own skins—was the most absurd of any I could have chosen. Yes, Grandmother was right. I don't know how "clever" I was, but my stupidity certainly exceeded all bounds.

6

• *My last year*

The last year of my former existence, which came to an end in February 1937, was very confused, but it was perfectly clear that I was heading straight for disaster, particularly after the Zinovyev-Kamenev trial, the Kemerovo affair, the trial of Pyatakov and Radek.*

The news burned, stung, clawed at one's heart. After each trial, the screw was turned tighter. The hideous term "enemy of the people" came into use. Every region and every national republic was obliged by some lunatic logic to have its own crop of enemies so as not to lag behind the others, for all the world as though it were a campaign for deliveries of grain or milk.

I myself was marked, I felt, and I wasn't allowed to forget it for a single moment. I spent almost the whole of that year in Moscow, since my appeal had been referred to the Party control commission and I therefore had to make endless visits to its office in Ilyinka Street. But as my husband was still a member of the Central Executive Committee of the USSR, I had a comfortable room at the Moskva Hotel, and on my frequent trips from Kazan I was met and taken to the station by a car from the Moscow office of the Tartar Republic. They also took me to and from Ilyinka Street where my fate was being decided. Such were the paradoxes and incongruities of the time.

That summer Gorky died, and at his funeral I saw Stalin for the first and last time in my life. I walked in the ranks of mourners from the Writers' Union, so I got a close look at him.

It would be an exaggeration if I were to ascribe to myself now any specially profound thoughts I had on that occasion about his role in the tragedy which was soon to engulf our Party and our country. But it would not be

---

* Grigory Zinovyev, 1883–1936, Lev Kamenev, 1883–1936, Grigory Pyatakov, 1890–1937, and Karl Radek, 1885–1939, were all prominent Bolshevik leaders. The Kemerovo affair was one of the many cases of alleged industrial sabotage.

untrue to say that I looked at his face without any sense
of hero worship and that it struck me by its ugliness—it
was not at all like the majestic countenance which gazed
benignly on us from millions of portraits. Not only did I
look at him without adulation—I even had a feeling of
suppressed hostility, though as yet only half conscious and
born of instinct rather than reason.

And to think of what was going on around me! Fyodor
Gladkov,* by then an old man, was walking alongside me,
and every time he looked at Stalin an expression of religious
fervor came into his face. And on my other side was a
young woman writer from Vologda. I remember the awe
and ecstasy with which she murmured: "I have seen Stalin.
Now I can die. . . ." I remember, too, my exasperation
and the words that flashed into my mind: "You idiot girl!"

Evidently some sixth sense told me that this man was to
be the evil genius in my life and that of my children. At
all events, when Makarovsky, the deputy director of the
schools department of the Central Committee and a good
friend of mine, once asked me if I would like sometime to
have a word about my case with the "boss," I was appalled.
"No, no, at least let him not know me personally." The
naïve monarchistic notion of the kindly ruler ignorant of
the abuses perpetrated by his officials was not one with
which I felt any sympathy even at that early stage of my
long and steep road. Whether Makarovsky himself, when
he was jailed, realized how right I had been about this,
I don't know.

The most varied characters turned up among the
"marked" people who besieged the Ilyinka Street corridors.
Women wept and men cursed; some waited meekly to

* Prominent Soviet writer, 1883–1964, best known for his novel *Cement*.

learn their fate, others vented their fury on the investigators. I remember a conversation with the manager of a Kharkov factory, sitting disconsolately on the hard wooden seat next to me.

"Cigarette?" He held out his case.

"No, thank you. I don't smoke."

"Really? How can anyone in our position manage without it? What do you do when . . ." He tapped his chest as though at a loss for words. "What do you do to soothe your feelings?"

"I go to the theater. Every day. Okhlopkov yesterday, today the Maly." *

"Does it help?"

"Up to a point, yes."

A working man of about forty, with pleasant brown eyes and a soft ingenuous mouth, broke into our conversation:

"You comrades can still make light of it. I can't. . . . My wife's crippled with rheumatism, hasn't walked for over a year, and we've got three children. I've been chucked out of the Party, and theaters don't do me much good. I come from Zaporozhye, and now my money's running out. I'm a compositor. Been in the printing business all my life."

The factory manager held out his case.

"Here, have a smoke, old man. You mustn't mind our jokes—it's just our funny sense of humor. What's brought you here?"

The other said nothing for a moment, then he leaned forward as though weighed down by an intolerable bur-

---

* Nikolay Okhlopkov, 1902–1966, theater director; Maly, one of Moscow's main theaters.

den, slapped the sides of his ancient knee-high boots, and cried in a desperate voice:

"Plekhanov's * been the ruin of me."

"What d'you mean?"

"We had this political study circle, you see. Party doctrine and all that. We had to learn about the new type of Party, and I—well, it's just too bad, but I didn't have the time. I had my wife ill in bed and the children to look after, and I had no time for anything, never mind political studies. Once they asked me, 'Who founded the new type of Party?' Well, all I need have said was, 'Look, I'm terribly sorry, but I haven't done my homework, I've got so many troubles at home that I didn't have a chance to open a book.' But I thought I heard somebody behind me whisper 'Plekhanov.' So, like a damned fool, I blurted out 'Plekhanov'—and that was what finished me. First I got a reprimand, and then things went from bad to worse. They started calling me a Menshevik, of all things! It seems there used to be a lot of Mensheviks in the printing trade, and they figured I was 'contaminated.' So they threw me out of the Party and I lost my job, and the children are starving. And my wife . . ."

His face worked with pain.

"She can't stand much more. She'll die. . . ." He paused again, then added: "And all because of Plekhanov."

Soon after, he was called in for interrogation, and through the door we heard the question put to him:

"Do you admit yourself guilty of having used the meetings of a political study circle in order to spread Menshevik, anti-Party views?"

---

* Georgy Plekhanov, 1857-1918, prominent Marxist theoretician, associated with the Mensheviks. Though he was never formally denounced, undue interest in him had become "unhealthy" by 1937.

At one point it seemed that things had taken a turn for the better on Ilyinka Street. Sidorov, a member of the Party board and a political commissar, heard me out with sympathy and became indignant over a passage in Beylin's report which said among other things that I was to be forbidden to conduct Marxist-Leninist propaganda.

"Damn it all, this is really too much! How can you forbid a Communist to carry on Marxist propaganda? The fellow's lost all sense of proportion."

He assured me that the penalty would be reduced. And indeed, in November I was notified that the "severe reprimand and warning" imposed on me by the Party board of the Tartar Republic had been amended to a "severe reprimand." Moreover, the ban on my engaging in propaganda work was lifted, and the charge of tolerance toward anti-Party elements was reduced to "relaxation of vigilance."

"In a year or so, when things quiet down a bit, we'll see about removing the sentence altogether," Sidorov said consolingly as he took leave of me, and his obvious sincerity made it clear that this level-headed man with his long record of work for the Party really did believe that things might "quiet down." Not even such experienced Party members as Sidorov could foresee the scale of the coming events. So it was not to be wondered at that I, the happy possessor of a severe reprimand *without* warning, immediately went back to Kazan with my mind almost entirely at rest.

But alas, my illusions were soon shattered. I had literally not finished unpacking when a telegram arrived from the Party control board: "Please come to Moscow at once for fresh hearing of your case. Yemelyan Yaroslavsky."

Later I learned that Beylin, who had turned up in Moscow at the very time when my sentence was reduced and who could not tolerate the blow to his self-esteem, had immediately gone to Yaroslavsky to complain about Sidorov and protest against the reversal of his, Beylin's, decision. He had also thrown in some additional charges against me. I was guilty, it seems, of associating not only with Elvov but also with Mikhail Korbut, "persons already condemned."

This time again Grandmother said to me:

"Don't you go to Moscow, Genia, whatever you do. Why not slip away to Pokrovskoye?"

And again I replied:

"How can I, Grandmother? How can a Communist run away from the Party?"

So I left again for Moscow that very evening and saw Yaroslavsky, who accused me of "not denouncing" Elvov's erroneous article—he who had *himself* included the article in the four-volume *History of the All-Union Communist Party*, of which he was the editor. It was enough to make one's head spin.

7

• *Life counted in minutes*

From that moment, events rushed on with breath-taking speed. I spent the two and a half months until my arrest in tormented conflict between reason and the kind of foreboding which Lermontov called "prophetic anguish."

My mind told me that there was absolutely nothing for which I could be arrested. It was true, of course, that in the monstrous accusations which the newspapers daily hurled at "enemies of the people" there was something clearly exaggerated, not quite real. All the same, I thought to myself, there must be something in it, however little— they must at least have voted the wrong way on some occasion or other. I, on the other hand, had never belonged to the opposition, nor had I ever had the slightest doubt as to the rightness of the Party line.

"If they arrested people like you they'd have to lock up the whole Party," my husband encouraged me in my line of reasoning.

Yet, in spite of all these rational arguments, I could not shake off a feeling of approaching disaster. I seemed to be at the center of an iron ring which was all the time contracting and would soon crush me.

The journey back to Moscow at Yaroslavsky's summons was terrible. I was very near to suicide.

I shared the first-class compartment with a woman doctor I knew, a children's doctor, Makarova, who had been to Kazan to defend her university thesis. She was a pleasant woman, not given to many words, gentle in her movements and with an attentive, thoughtful expression.

I thought that by talking about trifles I had more or less succeeded in hiding the state I was in. But suddenly— it had nothing to do with our conversation—she stroked my hand and said quietly:

"I'm very sorry for all the Communists I know. It's a very hard time for you. Anyone can be accused."

That night I was overwhelmed by such unspeakable misery that, as quietly as I could, I stole out of the compartment and along the corridor to the platform at the end of the carriage. My mind was empty of thought, except for some verses by Nekrasov * which fitted themselves to the rhythm of the train. Their theme was that those whose lives had been broken could still assert their courage by dying. The beat was hammered out by the wheels and the pulse throbbing in my temples. It was to escape from it that I had left the compartment. The November wind blew open my light dressing gown and for a moment distracted me. Then my wretchedness beset me again.

I opened the carriage door slightly. The cold air rushed in my face. I looked down into the noisy blackness, my mind finally invaded by a torturing vision. One step . . . one instant . . . and I wouldn't have to go to Yaroslavsky. I would never need to be afraid of anything.

I felt a strong but gentle grip on my arm. Makarova should have been a psychiatrist. Without any exclamations, without showering me with words, she led me back to the compartment and made me lie down. All she said as she stroked my hair was:

"All this will be over one day, and you've only got one life."

I would never have thought that Yaroslavsky, who was known as the "conscience of the Party," could have woven such a web of lying syllogisms. It was he who first explained to me the theory which became popular in 1937, that "when you get down to it, there is no difference between 'subjective' and 'objective.'" Whether you had committed a crime or, out of inadvertence or lack of vigilance, "added grist" to the criminal's mill, you were equally

* Nikolay Nekrasov, 1821–1877, author of the famous epic poem "Who lives well in Russia"; known for his vigorous denunciation of the evils of Tsarist Russia.

guilty. Even if you had not the slightest idea of what was going on, it was the same. The chain of "logical" reasoning in my case was as follows: "Elvov's article contained theoretical errors. Whether he intended them or not is beside the point. You who worked with him and knew he had written the article failed to denounce him. This is collusion with the enemy."

The charge of "relaxation of vigilance," formulated by the kindly and conscientious Sidorov, was replaced by one even harsher than Beylin's: Yaroslavsky accused me of "collaborating with enemies of the people."

So the i's were dotted—"collaboration with the enemy" was a specific, punishable, criminal offense.

My composure left me. I shouted and stamped my feet at that venerable old man. I might have gone for him with my fists if the wide, glossy surface of the desk had not been between us. I no longer remember exactly what I said to him, but it amounted to a counter-accusation. Yes, I was driven to such despair that I asked him plain questions, dictated by common sense. This was the height of bad form, for we were all supposed to pretend that syllogisms invented by sadists reflected the normal processes of the human mind. If anyone asked a question which showed up this lunacy for what it was, his hearers were either outraged or smiled condescendingly and treated him as an idiot.

But in my state of excitement that day in Yaroslavsky's office I actually shouted at him:

"All right, so I didn't denounce Elvov's article. But you—you not only didn't denounce it, you edited it and published it in your four-volume history of the Party. Why are you judging me and not I you? I'm thirty, you're sixty. I am a young member of the Party, while you are its 'conscience.' Why must I be torn to pieces and you left sitting behind that desk? It's a disgrace!"

I saw a glimmer of fright in his eyes. He must have thought I was mad. How otherwise could I have dared to say such things in this room, a cross between a shrine and a court of justice? But he at once resumed his stern mask of bigoted righteousness. He said with an almost natural tremor in his voice:

"No one knows my mistakes better than I do. Yes, I have wronged the Party—I, a man whose very existence is unthinkable outside its ranks."

It was on the tip of my tongue to ask another insanely daring question:

"Why can you redeem your mistake by merely being conscious of it, while I must pay for mine with my blood, my life, my children?"

But I didn't say it. My excitement was over. Instead, I felt horror. What had I said? What would they do to me now? Then the horror too died down and was replaced by a merciless clarity of vision: nothing would make any difference, nothing was any use. The time had come either to die silently or to tread the path to Golgotha along with others, thousands of others. When I was finally told to go back to Kazan where I would soon be notified of the decision, I could not wait to begin my journey. I knew for certain now that what was left of my life could not be measured in years or months but in minutes and I must hurry back to my children. What would become of them?

8

• The year 1937 begins

So it began—that accursed year which left its mark on the lives of millions. I saw it in—the last New Year of my old life—at Astafyevo, the rest home of the Central Executive Committee of the USSR, near Podolsk, not far from Moscow.

On my return to Kazan from my interview with Yaroslavsky, I found my elder son Alyosha seriously ill with malaria. The doctors advised a change of air. The school holidays were just beginning so I could take him away. My husband was able to get quarters for us in Astafyevo. He was pleased for my sake that I was going away again:

"The less you are seen in Kazan at the moment, the better."

He too was racked by anxiety at that time. The arrests were in full swing. Several people whom we knew well had disappeared. One of the first was Professor Aksyantsev, an old Party member and the head of the Tuberculosis Institute. The next to go was the rector of the university, Vekslin, whose selfless devotion to the Party was proverbial. I remember him in his shabby old overcoat—a man who had fought on every front all through the civil war, a hero of the siege of Perekop.

My husband took to spending more time at home. He was worn out by the regional committee sessions at which he, as a member of the board, had to listen in silence to endless hair-splitting discussions of the Elvov case, and of me and my part in it.

He was unused to staying at home in the evenings, and paced silently up and down the room, stopping from time to time to say:

"Well, how can you be sure about Vekslin? He's an impetuous sort of person. Perhaps he really was mixed up in something or other. . . ."

He began to take more notice of the children—he used only to play games with them before. He even noticed that Vasya's coat was shabby and he needed a new one.

But if I ever tried to talk to him frankly about what was going on, he immediately adopted the orthodox attitude. Of course he trusted me unreservedly and knew that I was innocent. But, as a member of the regional committee, he could not bring himself to share the view of the situation which was more and more clearly taking shape in my mind. It suited him better to think that a mistake had been made in my own particular case. He behaved chivalrously at the meetings, at which, again and again, he was called on to "dissociate" himself from his wife. There he said plainly that he knew I was a loyal Communist. But at home sometimes . . .

"Paul darling, what do you think really is happening in the Party?"

"It's hard to tell, Genia, I know. But what can one do? I suppose it's a passing phase. . . ."

"A passing phase to what? To prison for all Party members?" My bitter humor angered him.

"I'm sorry, Genia, that's the sort of joke that leaves a bad taste. You must forget your own wrongs. You mustn't bear a grudge against the Party."

Sometimes we had serious quarrels. I remember one unhappy scene late at night, in the deserted Lyadsky Park opposite our house. We had gone for a stroll before bedtime. We were talking casually but, almost against our will, we kept reverting to the same subject. I said something angry and mocking about Yaroslavsky. My husband burst out:

"What on earth do you mean, talking like that! You'll land us both in jail before you know it."

"Don't worry, we'll end up there anyway. . . ."

I pulled my arm away from his. Frightened by his own words, he tried to keep hold of it, but I struggled so hard that my gold wrist watch fell off and buried itself in a snowdrift beside the path. We searched for over an hour, but in vain.

Our troubled faces, as we bent over the snowdrift and dug into it with bare hands, seemed to foreshadow the misfortune that was almost upon us. The dreadful thing was that each of us knew exactly what the other was thinking: that we were only pretending to be upset by the loss of the watch and determined to find it. It wasn't the watch but our lives that were lost. We also knew that we were only pretending to be upset by the quarrel and to think it important. What could a quarrel between husband and wife matter *now?* We no longer had a part in life with its ordinary human relations. But this was only between the lines, as it were, and was unspoken even between ourselves.

Astafyevo, the former estate of Prince Vyazemsky, with its memories of Pushkin, was now a sort of super-Livadia for Moscow people. During the winter holidays it was filled by a large number of "ruling class" children who divided all those around them into categories according to the make of their cars. Lincolns and Buicks rated high, Fords low. Ours was a Ford, and Alyosha was quick to sense the difference this made.

"They're mean, Mother," he said. "You should hear what they say about their teachers."

Although the food was as good as in the best restaurants

and each room had its bowl of fruit, replenished as often as necessary, some of the ladies you met in the lounge turned up their noses at the cooking and compared it un-favorably with other places they knew.

The holiday was a real "Masque of the Red Death." Had they but known it, nine tenths of these people in Astafyevo were doomed, and within months nearly all of them were to exchange their comfortable suites for bunks at the Butyrki prison. Their children, who knew so much about the makes of cars, were packed into special children's homes, and even their chauffeurs were pulled in for "complicity" in something or other.

On New Year's Eve a lavish dinner was served in the dining room. The ladies put on their finery, and Alyosha insisted that I too should wear a new dress.

"You can't go in your old one, Mother. I do so like it when you look nice."

At five to twelve, just as our glasses were filled, I was called to the telephone. I hurried out happily, thinking it was my husband. The Central Executive Committee was to meet at the beginning of January and I thought he must already be in Moscow and was calling to wish me a happy New Year.

But the deep voice at the other end of the line was that of a chance acquaintance whom I didn't even like very much, and who had taken it into his head to offer his good wishes. By the time I got back to the party the New Year was in. The twelfth stroke of the gong sounded as I walked into the dining room. Alyosha had moved away and was clinking glasses with someone else. When he turned to me, two minutes of 1937 had gone by.

I didn't see it in with Alyosha, and it was the last we ever spent together.

9

At the beginning of February we returned to Kazan, and
I at once received a summons to the district committee.
Why Yaroslavsky had decided to refer my case back to
Kazan, I don't know. Perhaps, after the insolent things I
had said to him, he preferred not to set eyes on me again.
But more probably there had been a general decision to
have cases of expulsion dealt with on a lower level. There
were more and more of them every day, and the control
commission had more work than it could cope with.

My summons was for February 7th. The secretary of
the committee, Biktashev, had been a student of mine at
the Tartar Communist University. His face was a picture
of agony as he listened to the charges. I only vaguely remember what they were on this occasion, or how they
differed from the last Moscow version. I was hardly listening. Both the members of the board and I knew that my
expulsion was a foregone conclusion, and we both wanted
to get the wretched business over as quickly as possible.

"Any questions?"

"No."

"Any comments?"

"No."

"Perhaps there's something you'd like to say?" Biktashev
asked me in a hoarse voice, without raising his eyes from
the file in front of him. I could see he was afraid that I
would. As if it weren't obvious that he himself was suffering and could do nothing.

But everything was clear to me. I no longer wished to talk about anything. Quietly going to the door, I said in a whisper:

"Settle it without me. . . ."

We all knew that this was a break of Party regulations— no decision concerning a member could be taken in his or her absence—but what were regulations now? All Biktashev remembered was to say at the last moment:

"Your Party card . . . You have it with you?" Clearing his throat, he added:

"Leave it here. . . ."

There was a pause, and now Biktashev and I looked each other in the eye. The same memories must have come to both of us—of that time ten years ago when I, a young lecturer, new to my job, was tutoring him, a half-literate boy newly arrived from his village. That the boy had risen to be secretary to the district committee was in part my work. To think of all the difficulties and little triumphs we had gone through together, and how many times I had corrected his homework! And how cheerful and eager his narrow Mongol eyes had been, and how abashed and sad they were now. . . .

All this went through my mind and, I felt sure, through his as well. His voice trembled as he said again:

"Leave the card here . . . for the time being. . . ."

"For the time being" was a poor effort to comfort me, to give me hope, as though he was trying to say that I would get it back someday, and that these things couldn't go on for ever.

I felt sorry for my old pupil Biktashev, a good, hardworking boy. It was worse for him than for me. Some actors in the horror play had been cast as victims, others as persecutors, and these were the worse off. At least my conscience was clear.

"Yes, I have my card."

It still looked quite new: there had been a new issue throughout the country as recently as 1935. What care I had taken of it, how afraid I had been of losing it! I put it down on the desk.

My husband was waiting in the street outside. We set off on foot. It would not have done to go by streetcar, looking as upset as we did. We said nothing until halfway home, then he asked:

"Well, what happened?"

"I had to leave my card. . . ."

He sighed softly. It was clear to him now how close we were to the precipice.

10

• *That day*

After my expulsion from the Party, eight days went by before my arrest. All those days I sat at home, shut in my room and not answering the telephone. I was waiting; so were all my family. What for? We told each other that we were waiting for the leave my husband had been promised at this unusual time. When it came we would go again to Moscow and try to get people to help me. We would ask Razumov, who was a member of the Central Committee, to help us.

In our hearts we knew perfectly well that none of this would happen, that we were waiting for something quite different. My mother and my husband took turns watching over me. My mother sometimes fried some potatoes. "Do have some, darling. Remember how you liked them done this way when you were little?" Every time my husband had been out and came home he rang in a special way and shouted through the door:

"It's only me, open up!"

It was as if he were saying:

"It's still only me, not them."

We started a purge of our books. Our old nurse carried out pail after pailful of ashes. We burned Radek's *Portraits and Pamphlets*, the *History of Western Europe* by Friedland and Slutsky, Bukharin's *Political Economy*.

The "Index" grew longer and longer, and the scale of our *auto da fé* grander and grander. We even had to burn Stalin's *On the Opposition*. This too had become illegal under the new dispensation.

A few days before my arrest, Biktagirov, second secretary of the Party municipal committee, was summarily removed from a meeting at which he was presiding. His secretary came in:

"Comrade Biktagirov, you are wanted."

"In the middle of a meeting? What nonsense. Tell them I'm busy."

But the secretary came back:

"They insist."

So he went, and was invited to put on his coat and go "for a short drive."

My husband was even more puzzled and shaken by this event than by my own expulsion from the Party. A secretary of the municipal committee! And he too had "turned out to be an enemy of the people" . . .

"Really, the Cheka * is getting a bit above itself. They'll have to let a good many of these people out again."

He was trying to convince himself that it was all no more than a checkup, or a misunderstanding of some sort, temporary and almost ludicrous. Surely on our next free day we would find Biktagirov sitting once more at table at Livadia, telling with a smile how he had almost been mistaken for an enemy of the people.

But the nights were very unpleasant. The windows of our bedroom faced the street and cars drove past all the time. And how we listened in fear and trembling when it seemed as though one of them might be pulling up in front of our house. At night, even my husband's optimism would give way to terror—the great terror that gripped our whole country by the throat.

"Paul! A car!"

"Well, what of it, darling? It's a big town, there are plenty of cars."

"It's stopped. I'm sure it has."

My husband, barefooted, would leap across to the window. He was pale but spoke with exaggerated calm:

"There, you see, it's only a truck."

"Don't they use trucks sometimes?"

We would fall asleep only after six, and when we woke there was the latest news about who had been "exposed."

---

* Cheka: abbreviation of Russian for "Extraordinary Commission for the struggle against counter-revolution and sabotage," as the Soviet security service was called when it was first created by Lenin. Cheka remains as a colloquial term for the Secret Police.

"Have you heard? Petrov has turned out to be an enemy of the people! How cunning he must have been to get away with it for so long."

This meant that Petrov had been arrested overnight.

Sheaves of newspapers would arrive, and by now there was no telling which was the *Literary Gazette* and which, say, *Soviet Art*. They all ranted and raved in the same way about enemies, conspiracies, shootings. . . .

The nights were terrifying. But what we were waiting for actually happened in the daytime.

We were in the dining room, my husband, Alyosha, and I. My stepdaughter Mayka was out skating. Vasya was in the nursery. I was ironing some laundry. I often felt like doing manual work; it distracted me from my thoughts. Alyosha was having breakfast, and my husband was reading a story by Valeria Gerasimova aloud to him. Suddenly the telephone rang. It sounded as shrill as on that day in December 1934.

For a few moments, none of us picked it up. We hated telephone calls in those days. Then my husband said in that unnaturally calm voice he so often used now:

"It must be Lukovnikov. I asked him to call."

He took the receiver, listened, went as white as a sheet, and said even more quietly:

"It's for you, Genia. Vevers, of the NKVD."

Vevers, the head of the NKVD department for special political affairs, could not have been more amiable and charming. His voice burbled on like a brook in spring.

"Good morning, dear comrade. Tell me, how are you fixed for time today?"

"I'm always free now. Why?"

"Oh, dear, always free, how depressing. Never mind, these things will pass. So anyway, you'd have time to come and see me for a moment. The thing is, we'd like some information about that fellow Elvov . . . some additional information. My word, he did land you in a mess, didn't he! Oh, well, we'll soon sort it all out."

"When shall I come?"

"Whenever it suits you best. Now, if you like, or if it's more convenient, after lunch."

"How long is it likely to take?"

"Oh, say forty minutes, perhaps an hour."

My husband, who was standing beside me and could hear, was making signs and whispering to me to go at once, so that Vevers shouldn't think I was afraid—there was nothing to be afraid of.

I told Vevers I'd go at once.

"Perhaps I'll just stop at Mother's on the way," I said to my husband.

"No, don't. Go at once. The sooner it's all cleared up, the better."

He helped me to get quickly into my things. I sent Alyosha off to the skating rink. He went without saying good-by. I never saw him again.

For some strange reason, little Vasya, who was used to my going and coming and always took it perfectly calmly, ran out into the hall after me and kept asking insistently:

"Where are you going, Mother, where? Tell me. I don't want you to go!"

But I could not so much as look at the children or kiss them—if I had, I would have died then and there. I turned away and called out to the nurse:

"Fima, do take him. I haven't time for him now."

Perhaps it was just as well not to see my mother either. What must be must be, and there's no point in trying to postpone it. The door banged shut. I still remember the sound. That was all . . . I was never again to open that door behind which I had lived with my dear children.

On the stairs we met Mayka, back from the skating rink. She was a child who understood everything intuitively. She said nothing, and didn't seem to wonder where we could be going at that unusual hour. Her enormous blue eyes wide open, she pressed herself against the wall, and so deep an understanding of pain and horror showed in her twelve-year-old face that I dreamed of it for years afterward.

Our old nurse Fima caught up with us at the front door. She had run down to tell me something. But she looked at me and said nothing. She only made the sign of the cross after us as we moved away.

"Let's walk, shall we."

"Yes, let's, while we still can."

"Don't be silly. That's not the way they arrest people. They want some information, that's all."

We walked for a long time in silence. It was a lovely, bright February day. Snow had fallen that morning and was still very clean.

"It's our last walk together, Paul darling. I'm a state criminal now."

"Don't talk nonsense, Genia. I told you before, if they arrested people like you they'd have to lock up the whole Party."

"I sometimes have the crazy idea that that's what they mean to do."

I waited for my husband's usual reaction, thinking he would scold me for my blasphemous words. But instead, he himself gave way to "heresy" and said he was sure of the innocence of many of those who had been arrested as enemies, and he talked indignantly about very highly placed people indeed.

I was glad we were once again of one mind. I imagined then that everything was quite clear to me, though in fact many bitter discoveries still lay before me.

But here we were at the well-known address in Black Lake Street.

"Well, Genia, we'll expect you home for lunch."

How pathetic he looked, all of a sudden, how his lips trembled! I thought of his assured, masterful tone in the old days, the tone of an old Communist, an experienced Party worker.

"Good-by, Paul dear. We've had a good life together."

I didn't even say "Look after the children." I knew he would not be able to take care of them. He was again trying to comfort me with commonplaces—I could no longer catch what he was saying. I walked quickly toward the reception room, and suddenly heard his broken cry:

"Genia!"

He had the haunted look of a baited animal, of a harried and exhausted human being—it was a look I was to see again and again, *there.*

11

I opened the door briskly, with the boldness of despair. If you are to jump over a cliff, better take a run at it and not pause on the brink to look back at the lovely world you are leaving behind.

But the leisurely bureaucratic manner in which they made out my pass—the time of my arrival duly entered and a space left blank for the time I would leave—all this gave me a flash of hope: perhaps they really only wanted to question me about Elvov.

I went up to the first floor, then the second. People were walking busily about the corridors, and from behind a glass door came the rattle of typewriters. A young man I had seen somewhere before actually nodded with a casual "Good morning!" It was an office just like any other!

By now completely composed, I climbed the third flight of stairs and paused for a moment outside Room 47. I knocked and, not quite catching the answer, walked in—to be brought up short by Vevers's glance. I looked straight into his eyes.

They were eyes which would have been worth showing in close-up at the cinema. They were naked. They made not the slightest attempt to conceal their cynicism, cruelty, and anticipation of the pleasure of torturing a victim. No commentary was needed—they spoke for themselves.

But I did not give up immediately. I tried to make it plain that I still considered myself a human being, a Com-

munist, and a woman. I said how do you do politely and,
surprised that he did not offer me a chair, I asked:

"May I sit down?"

"Sit if you're tired," he grunted contemptuously. His
face was contorted by the same expression—a mixture of
hatred, scorn, and mockery—which I was to see hundreds
of times on the faces of his fellow apparatchiks and of
heads of prisons and camps.

I learned later that this grimace was part of the interrogators' stock in trade and that they were made to practice
it before a looking glass. But seeing it for the first time,
I felt sure that it expressed Vevers's own attitude to me
personally.

For a few minutes we sat in silence. Then he took a
blank sheet of paper and wrote on it slowly, in large letters
so that I could read it, the heading "Record of Interrogations" and under it my husband's surname. I corrected him,
giving my maiden name.

"Trying to protect him? It won't help."

He raised his eyes to me again. This time they were
gray and bleary with boredom.

"Well, how do you stand with the Party?"

"Surely you know. I've been expelled from it."

"So I should hope! You don't expect us to keep traitors
in, do you?"

"Why do you insult me?"

"Insult you? Why, death would be too good for you!
You turncoat! You agent of international imperialism!"

Surely he must be joking, he couldn't mean such things.
But he did. Working himself up more and more, he shouted
across the room, pouring invective at me. True, his invective was only political. It was still only February 1937. By

June, he would be treating prisoners to the choicest gutter oaths.

He ended a long tirade with a blow of his fist on the desk, so that the glass top reverberated. To its plaintive whine, he thundered the words which served, as it were, as the grand finale of my two years' ordeal:

"I hope you realize that you're under arrest."

The green and gold pattern on the wallpaper of Vevers's office spun around, the room lurched.

"It's illegal! I have committed no crime." I mumbled, my tongue dry.

"What! Illegal? And what's this? It's a warrant for your arrest, signed by the public prosecutor. It's dated the fifth of February. Today is the fifteenth. We didn't have time to get around to you earlier. Only today I was telephoned from a certain quarter, to know why we were letting enemies of the people walk about scot-free."

I got up and took a step toward the telephone.

"Let me tell my family."

He burst out laughing.

"You're a cool one, you are! Since when are prisoners allowed to make phone calls?"

"Then will you call them yourself?"

"All in good time. Aksyonov doesn't care all that much, you know. He's disowned you already. Pretty funny, when you come to think of it—a member of the government and of the regional committee board, with a wife like you! But that isn't the point now. We've got to draw up the record. Answer my questions." He wrote something, then read out:

"It is known to the investigators that you belonged to a secret terrorist organization among the editorial staff of Red Tartary. Do you admit this?"

"Nonsense. There was no such organization, and I didn't belong to anything of the sort."

"Shut up!"

He banged the desk again and the glass rang out plaintively again.

"You can drop that snooty tone with me, it's no use acting the lady any more. It's prison bars for you now."

"What right have you to shout and bang the table at me? I demand to see the head of your department, Comrade Rud."

"Oh, you demand, do you? I'll teach you to demand things."

He pressed a buzzer. A woman in the uniform of a prison wardress came in.

"Search her!"

I was still very inexperienced. All I knew of such things was what I had read in the memoirs of old Bolsheviks or in books about the "People's Will." * So it was with surprise as well as disgust that I watched the movements of those shameless hands as they rummaged in my pockets and crawled over my body like snails.

The search was over. My terrorist equipment consisted of a pair of nail scissors I happened to have in my bag.

Captain Vevers pressed another button and a warder came in—this time a man. Vevers stared at me again, his leaden eyes full of hatred and contempt.

"And now off to the cellar with you! And there you'll sit till you confess and sign everything."

---

* Populist organization which in the 1870's and 1880's engaged in terrorist activities against the Tsarist regime.

12

• *The cellars at "Black Lake"*

"Black Lake" was, strictly speaking, the name of one of Kazan's municipal parks. Before the Revolution it had been a favorite haunt of shopkeepers out on a spree and there had been an expensive restaurant and a vaudeville theater on the grounds. Nowadays, these were used for exhibitions, and in winter there was a skating rink.

But after the NKVD had moved into a building on Black Lake Street, the name lost its original association and took on the same meaning as "Lubyanka" in Moscow. People said: "Watch your tongue, or you'll find yourself at Black Lake," or "Have you heard? They took him off to Black Lake last night."

The very words "cellars at Black Lake" aroused terror. And here I was, under guard, on my way to those cellars. How many steps down—a hundred, a thousand? I don't remember. I remember only that my heart sank at each step, and I had the sardonic thought that this must be what sinners feel when they see hell with their own eyes after having said the word unthinkingly so many times while they were still alive.

A heavy metal door groaned. It led only to the ante-room, a smoky, windowless room lit by a small ceiling lamp. A pale-faced warder sat behind a desk. He had heavy pouches under his eyes, and the eyes were as insultingly blank as those of a codfish. He looked at me as at an empty space. All that interested him was my watch. Being a woman, I wore no belt, and this was the point at which watches and belts were removed from newly arrived prisoners, so that they should not be able to tell the time or to hang themselves. He liked my watch—it was a handsome foreign one my husband had given me when I lost my old one in the snowdrift. As he examined it approvingly, a spark of life came into his glazed fish eyes. Then he set about filling in a form—it seemed I needed one even to get into hell. Finally he exchanged a few words with my guard, apparently about which cell to put me in.

And here was hell itself. A second iron door led into a narrow passage dimly lit by a single bulb close to the ceiling. The bulb gave off a special prison light—a sort of dull-red glow. To my right was a gray, damp, blotched wall—it was hard to believe that this was part of the same building that housed Captain Vevers's well-appointed office. And on the left . . .

On the left was a long row of bolted and padlocked doors. The huge, rusty padlocks of antique design were such as merchants in pre-revolutionary days might have put on their stores in some out-of-the-way province. Behind these doors were friends of mine, Communists who had been cast down into hell before me: Professor Aksyantsev, Biktagirov of the municipal committee, Vekslin, the university rector, and many others. . . .

The cold despair which filled me gave me an outward appearance of calm. I had mentally prepared myself for solitary confinement. So when a cell door, marked "3," opened with much creaking and grinding, and I saw the outline of a human figure inside, I took this as an unexpected gift of fate. I was not to be alone. This already was a blessing.

The door clanged shut, and as the sound of the warder's footsteps died away I said quietly to the young woman who stood by the wall:

"Hello, comrade."

For a few seconds she made no response. She was pretty and wore a smart, expensive sealskin coat which went well with her golden hair.

She gazed anxiously into my face, as though searching it for the answer to some question, then she smiled sadly and said:

"Hello. Sit here, on this stool. Don't take your coat off, it's cold. . . . You look so calm. I suppose you've no one left at home?"

"My children. . . . Two sons and a stepdaughter. My younger son is four and a half."

"How dreadful! Oh, you poor thing!"

She hurried over to where I sat, crouched in front of me, and looked searchingly into my eyes.

"What is it you want to ask me?"

She hesitated for a second, then took both my hands in hers and said, laughing:

"I'm not going to ask, I can see for myself. You're all right. I've been so afraid they'd put a spy in with me, who'd question me, and make me say something damaging to my father and me. Father is here too. So is my brother. Mother's all alone, she's an invalid, her hands were mutilated when the Chinese bandits raided us."

I still had no idea what she was talking about. It was all so far away from my small, closed world of Party intellectuals and scholars. Did she come from some border region? What was all this about Chinese bandits? But I felt a surge of friendliness in response to hers. No one with her pretty, open face and steady brown eyes could be wicked. I clasped the small hands she held out to me and said, trying to calm her:

"No, I'm not a spy. I won't ask you anything. You needn't tell me anything about yourself if you don't want to. But I'll tell you all about myself, so that you shan't be afraid. After all, we're together in this terrible trouble. . . . My name is Genia. Do call me that, even though you're younger than I am. What's yours?"

She was Lyama Shepel. Her real name was Lydia, but she had called herself Lyama as a child and it had stuck.

"I'm from the 'CFER.' "

"What's that?"

I knew of no such country or organization, but within a few months I was to meet dozens of people who described themselves in this outlandish way. The letters stood for "Chinese–Far Eastern Railway," and they were Russians, mostly skilled workers, who had served on this railway and come home after we sold it to the Manchurians in 1935. Many of them had spent several years in Manchuria, had come back full of patriotism, and were eager to work hard for their country. Nearly all had been arrested as spies, on the grotesque charge of having been "recruited" by the Japanese and Manchurian Secret Services.

Lyama was twenty-two. She had been through high school in Harbin and had worked as a typist at the railway office. Her father was an old railwayman. Her brother, a few years older than she, had got into trouble for Communist sympathies and moved to the Soviet Union, where he settled in the small industrial town of Zelenodolsk. Lyama and her parents had joined him a few months ago, after the railway was sold. A month ago, she, her father, and her brother had been arrested, and all three were charged with espionage.

"They made up a whole spy story. Poor Father has a weak heart, it'll be the death of him. And Mother can't earn her own living. As I told you, the bandits broke all her fingers."

When I tried to tell Lyama about myself, I found it was utterly beyond her. With all my teacher's training, I could not explain to this child of another world what exactly I was being accused of. All our talk of "lack of vigilance," "appeasement," "rotten liberalism," was so much Chinese to her, or rather it was sheer gibberish, for she did know quite a lot of Chinese. On the other hand, she was by now an experienced prisoner and lost no time in explaining to me all the ins and outs of the life ahead of me.

I noticed at this point that our two iron bunks were folded back against the wall and fixed to it with hooks. We were not allowed to lower them before the special signal given at eleven at night. Reveille was at six in the morning, and from then until the evening we could not lie down. We could only stand or sit on stools.

"It's the good warder who's on duty today. He always gives us more to eat. With the other, the cross-eyed one, you could starve to death. I expect today they'll let us go to the lavatory after supper—they've started at the other end of the passage."

Soon we heard the grinding noise of cell doors being unlocked, one after another. A smell of bad fish seeped into the cell, more sickening even than the all-pervading smell of mildew, slop pails, and excrement.

I looked on with amazement as my pretty, dainty cellmate devoured first her own portion of fish and dry oatmeal *kasha* and then mine as well.

"All right, I won't refuse yours today—I know nobody can eat it at first."

When she had finished she stood by the door, waiting for the order "Plates" and, as soon as the heavy door was pushed ajar, put our mess tins on the floor.

"We're forbidden to pass them anything from hand to hand."

After supper we were taken to the lavatory. We had to walk along the corridor in silence and in single file. The washroom was at the far end, so we passed all the cells. I stared hard at each door, as though I could see right through to the people inside.

Lyama carefully initiated me into the rules of this strange monastery and taught me the various ways of deceiving warders and interrogators.

"Quick, stand with your back to the door," she whispered as soon as the two of us were alone in the lavatory. In a flash she sprinkled the narrow wooden shelf over the washbasin with tooth powder and traced my initials.

"The next lot will read it, and if they're local people they may guess it's you. Then they can tell their neighbors. The main thing is to make contact."

"How will they pass it on to others?"

"By tapping it out on the wall."

"Do you know how to do that?"

"I'm learning. Our neighbor teaches me a little every day after dinner, when the guards are changed. I'll tell you what: when we go back, try to click your heels lightly, so that people can tell you're a woman by the way you walk, and when you pass Number 5, clear your throat. I believe some important Party member from Kazan is in there, and he might know your voice."

The window of our cell was protected not only by thick

bars but by a high wooden screen which left only a tiny scrap of sky to be seen above it.

"It's dark even by day. Too dark to read."

Lyama smiled at my naïveté.

"It's dark all right—in a cellar and with the window boarded up. But as for reading—don't worry, we're not allowed to anyway."

She dropped her voice to the faintest of whispers:

"Look carefully at the screen. Do you notice anything?"

No, decidedly, I didn't. Bars, boards—the world was locked away. But, as she showed me, there was after all a minute opening in it—a chink of light between the second and third boards.

"Let's hope they won't notice. Now that there are two of us, we can watch all through the exercise period and find out who the people in the cells are. The important thing is to establish contact—to discover who's being questioned about what, and what they replied. The interrogators are frightful liars, you know, you can't think what lies they tell."

Bursting to talk after her month in "solitary," Lyama went on till lights out, whispering excitedly about the treachery of the interrogators, and picnics in times past on the banks of the Chinese river Sungari, and the misery of seeing one's clothes go to ruin—especially her fur coat. It was genuine sealskin, you couldn't get one like it nowadays, and she had made a terrible mess of it, sitting in the damp cell and carrying slops and other filth.

I listened to her eagerly, full of sympathy and even of a strange feeling of shame before this girl from a world unknown to me. Although I was locked up in the same prison as she, I still felt I shared responsibility for the fact that the welcome this good, cheerful, intelligent Russian girl had received from the longed-for land of her fathers was the hospitality of this cell.

Lights out. Metal bunks, clanging as they were let down, filled the cellars with their infernal noise. I had a good look at my new bed. Still squeamish, I carefully spread my scarf and handkerchief over the gray pillow stuffed with straw.

"They may allow you to have parcels," Lyama tried to comfort me.

We went to bed. Lyama fell asleep with a happy smile on her face. Before she dropped off, she said:

"Good night, darling Genia! I'm so glad you're with me. I've just now realized how terrible it is to be alone for a whole month. Oh, I am a fool! Of course, I didn't mean I'm glad you were arrested—I'm only glad they put you in with me—you do understand?"

Yes, I understood everything. Or perhaps I had ceased to understand anything at all. The day had been too crowded with events. I closed my eyes and pictures drifted past me: the faces of my children, my husband's last hunted look, Vevers's shining eyes, the "codfish" goggling at my watch, and the Chinese river Sungari dotted with little boats and strange-looking people sailing in them. They were also from the CFER. How silly it sounded. Lyama . . . Only a few hours ago I had never heard of her existence, and now she was like a sister to me. . . .

I had almost sunk into the troubled and uncertain nightmare-ridden state that passes for sleep in prison. But I was not to sleep for long that first night. Again the rumbling and clanking of bolts and keys, and a bored stage whisper in my ear: "Get ready for interrogation."

13

I have often thought about the tragedy of those by whose agency the purge of 1937 was carried out. What a life they had! They were all sadists, of course. And only a handful found the courage to commit suicide.

Step by step, as they followed their routine directives, they traveled all the way from the human condition to that of beasts. Their faces, as time went by, defied description. I, at any rate, cannot find words to convey the expression on the faces of these un-men.

But all this happened only gradually. That night, Interrogator Livanov, who had summoned me, looked like any other civil servant with perhaps a little more than the usual liking for red tape. Everything about him confirmed this impression—the placid, well-fed face, the neat writing with which he filled the left-hand side (reserved for questions) of the record sheet in front of him, and his local, Kazan accent. Certain turns of speech, provincial and old-fashioned, reminded me of our nurse Fima and aroused a host of memories of home.

In that first moment I had a flash of hope that the madness might be over, that I had left it behind me, down there, with the grinding of padlocks and the pain-filled eyes of the golden-haired girl from the banks of the river Sungari. Here, it seemed, was the world of ordinary, normal people. Outside the window was the old familiar town with its clanging streetcars. The window had neither bars nor a wooden screen, but handsome net curtains. And the

plate with the remains of Livanov's supper had not been
left on the floor but stood on a small table in a corner of
the room.

He might be a perfectly decent man, this quiet official
who was slowly writing down my answers to his straightforward, insignificant questions: Where had I worked between this year and that, where and when had I met this
or that person . . . ? But now the first page had been
filled and he gave it to me to sign.

What was this? He had asked me how long I had known
Elvov, and I had answered, "Since 1932." But here it said,
"How long have you known the Trotskyist Elvov?" and
my reply was put down as "I have known the Trotskyist
Elvov since 1932."

"This isn't what I said."

He looked at me in amazement, as though it really were
only a question of getting the definition right.

"But he *is* a Trotskyist!"

"I don't know that."

"But we do. It's been established. The investigators have
conclusive evidence."

"But I can't confirm something I don't know. You can
ask me when I met *Professor* Elvov, but whether he is a
Trotskyist, and whether I knew him as a Trotskyist—that's
a different question."

"It's for me to ask the questions, if you please. You've
no right to dictate to me the form in which I should put
them. All you have to do is answer them."

"Then put the answers down exactly as I give them,
and not in your own words. In fact, why don't we have
a stenographer to put them down?"

These words, the height of naïveté, were greeted by
peals of laughter—not, of course, from Livanov but from

the embodiment of lunacy which had just entered the room in the person of State Security Lieutenant Tsarevsky.

"Well, well, what do I see! You're behind bars now, are you? And how long is it since you gave us a lecture at the club, on Dobrolyubov? * Eh? Remember?"

"Yes, I do. It was very silly, I agree. What could you possibly want with Dobrolyubov?"

My sarcasm was wasted on the lanky, tousled youth with the face of a maniac.

"So you want a stenographer! No more and no less! What a joke! Think you're back in your editorial office?"

With quick, jerky steps he went up to the desk, ran his eyes over the record, and looked up at me. His eyes, like Vevers's, were those of a sadist who took pleasure in his work, but there was also in them a lurking anxiety, a latent fear.

"So you're behind bars," he sneered again, in a tone of hatred as intense as though I had set fire to his house or murdered his child.

"You realize, of course," he went on more quietly, "that the regional committee has agreed to your arrest. Everything has come out. Elvov gave you away. That husband of yours, Aksyonov—he's been arrested too, and he's come clean. He's a Trotskyist too, of course."

I mentally compared this statement with Vevers's about Aksyonov disowning me. Yes, Lyama was right. They really were brazen liars.

"Is Elvov here, in this prison?"

"Yes! In the very cell next to yours. And he's confirmed all the evidence against you."

* Nikolay Dobrolyubov, 1836–1861, radical journalist and critic.

"Then confront me with him. I want to know what he said about me. Let him repeat it to my face."

"Like to see your friend, eh?" He added such a scurrilous obscenity that I could hardly believe my ears.

"How dare you! I demand to see the head of your department. This is a Soviet institution where people can't be treated like dirt."

"Enemies are not people. We're allowed to do what we like with them. People indeed!"

Again he roared with laughter. Then he screamed at me at the top of his voice, banged the table with his fist, exactly like Vevers, and told me I'd be shot if I didn't sign the record.

I noticed with amazement that Livanov, so quiet and polite, looked on unmoved. Obviously, he had seen it all before.

"Why do you allow this man to interfere with a case you're in charge of?" I asked.

His smile was almost gentle.

"Tsarevsky's right, you know. You can make things better for yourself if you honestly repent and make a clean breast of it. Stubbornness won't help you. The investigators have conclusive evidence."

"Of what?"

"Of your counter-revolutionary activity as a member of the secret organization headed by Elvov. You'd much better sign the record. If you do you'll be treated decently. We'll allow you to have parcels, and we'll let you see your children and your husband."

While Livanov spoke Tsarevsky kept quiet in readiness for his next attack. I thought he had just happened to come into Livanov's office. But after the two of them had been at me for three or four hours, I realized that it was part of a deliberate technique.

The blue light of a February dawn was casting its chill on the room by the time Tsarevsky rang for the warder. The same words ended the interrogation as those of Vevers the day before, but Tsarevsky's voice rose to a falsetto:

"Off to the cell with you! And there you'll sit until you sign."

Going down the stairs, I caught myself hurrying back to my cell. It seemed, after all, that I was better off there, in the human presence of a companion in misfortune—and the grinding of locks was better than the demented screams of un-men.

14

• *Stick and carrot*

By the end of a week I had learned the procedure so thoroughly that, as I walked in front of the warder to the interrogation room, I did not wait for his instruction but of my own accord turned right to Livanov's office, where I sometimes found Tsarevsky and sometimes the two of them waiting for me. So I was surprised when, as we got to the first floor, I heard the guard saying quietly but distinctly:

"Left!"

The new office was much better furnished than Livanov's. For some reason, the curtains over the large plate-glass windows had been drawn back, and I could not suppress a cry of astonishment and delight at seeing, as on a cinema screen, the skating rink in Black Lake Park. It was festively decorated with strings of colored lights, and I could see the brass band playing on its stand, and the dark figures of the skaters.

For a second I froze, incapable of tearing myself away. Was it possible that such things still existed in the world—a world in which there were punishment cells with standing room only, and "special methods" with which I was now daily threatened?

"Pretty, isn't it?" said a velvety baritone voice.

Only then did I notice a short, stocky man in military uniform, standing by the side window.

"It's a holiday today—Red Army Day. There's a big skating competition," he explained in the sort of voice he might have used at a tea party, adding cozily: "Your two eldest children must be there, Alyosha and Mayka—they both skate, don't they?"

Was it a hallucination? Had someone within these walls really spoken the names of my children? It was more than I could bear. I had promised myself again and again that "they" would never see me cry. But this was too unexpected, and my tears flowed in torrents.

"Oh dear . . . I'm so sorry I upset you. Do sit down—here in this armchair—you'll be more comfortable."

My companion was not in the least like "them." He reminded me rather of my bygone days at the university. His gray eyes were full of sympathy as he began a conversation which seemed to have nothing to do with my "case." He spoke of vocations in life: he was sure I had made a mistake in taking up teaching and research.

"You're a born writer. I was shown some of your newspaper articles yesterday."

I still had no idea what all this was leading to. But he
soon came to the point.

"An impulsive, emotional nature like yours—no wonder
you were taken in by the bogus romanticism of that
wretched underground."

Major Yelshin looked at me expectantly. But I had
learned a lot in the past week. I knew by now that however impassioned my protestations of innocence, they did
not convince anyone and were only likely to provoke
insults. I had realized that silence was golden, that I must
reply only to direct questions, and then as briefly as possible.

"Ah, well," the Major went on, "we've all been young,
we've all been carried away. Anyone can make a mistake."

Good Lord, did he really think I'd never read novels
and stories about the early days of the revolutionary movement? Wasn't there always a Tsarist policeman in them
who chided young terrorists in exactly these terms?

"Do you smoke?" He courteously offered me his cigarette case and went on, as though thinking aloud: "What
romanticism . . . Auguste Blanqui . . . Stepnyak-Kravchinsky and all that. . . . Do you remember *The House
on the Volga*?" *

I could see the Major was glad of a chance to display
the extent of his reading. Getting into his stride, he delivered a speech—it lasted about ten minutes—about how
mistaken my attitude was. I was not, after all, in the hands

* Louis Auguste Blanqui, 1805–1881, French utopian socialist.—Sergey
Stepnyak-Kravchinsky, 1851–1895, revolutionary terrorist and author
of several novels, including *The House on the Volga*.

of the Gestapo. There, it would have been right and fitting to be proudly silent, to refuse to sign records or to name my associates. But here I was in prison in my own country. He was sure that I was still a Communist at heart, in spite of my serious errors. I must disarm, go down on my knees to the Party, name those who had played on my impulsive, emotional nature and got me to join their corrupt secret society—and then go back to my children. They sent me their love, by the way. Only yesterday he had spoken on the telephone to Comrade Aksyonov. That loyal Communist was deeply pained to hear that his wife was compounding her mistakes by her wrong—to put it plainly—her un-Soviet behavior.

I sat silent, struck dumb, trying to look past him at the corner of the room. He thought I was looking at a plate of sandwiches on a small table in the corner.

"I'm sorry, I never thought of offering you any. Do help yourself. Are you hungry? You do look a bit pale. Though actually, it suits you. An attractive woman like you—no wonder Elvov lost his head! Not in the least surprising, is it?"

The pile of sandwiches with their thin slices of tender pink ham and imitation Swiss cheese was put in front of me.

Was I hungry? Sickened by the prison mess tins and the stinking fish, I had eaten almost nothing all that week except a chunk of black bread washed down with hot water.

"Thanks, I'm not hungry."

"Now, now, that's another bad thing. You look on us as enemies. You won't take food from us."

Again I said nothing, trying to look at neither the Major

nor the sandwiches. With a brief sigh, he took them away and replaced them by several sheets of paper and a fountain pen.

"Write it all down for us. All that happened, right from the beginning. I'll get on with my work, and you write it all down in as much detail as possible. Make it clear who the ringleaders were, and which of the people at the university and on the newspaper were particularly hostile to the Party line. Oh, and don't forget the Tartar writers either. But it's not for me to teach you how to write!"

"I'm afraid it's not in my line, Major."

"Why not?"

"Well, you yourself mentioned the kind of writing I do —articles, translations. But I've never tried my hand at detective novels, and I doubt if I could do the kind of fiction you want."

The Major smiled wryly but remained polite. Evidently his function was confined to the carrot, not the stick.

"You just write. We'll see how you get on."

"What's the point of writing about university people? Haven't they all been arrested by now?" I tried to draw out my affable host.

"Good heavens, no! Take Professor Kamay, for instance. Who would dream of arresting him? On what grounds? A former stevedore, a Tartar, a senior professor of chemistry, a devoted Party member."

"Yes, he must be the last professor who was once a stevedore. Now you're more inclined to make stevedores out of professors."

I had nothing to lose—by now I was convinced of this— so I allowed myself an occasional quip.

"Now, now," Major Yelshin said in fatherly rebuke, "doesn't it seem even to you that your little joke has a whiff of Trotskyism about it? Doesn't it come straight out of the Trotskyist arsenal?"

I might as well use the pen and paper for something, I told myself. So I wrote. For hours on end I wrote a statement to the local head of the NKVD, whom I had not seen since my arrest but had once talked to at a meeting of Party activists. I wrote about illegal methods of investigation, about threats and sleepless nights, about Tsarevsky and Vevers. I asked to be confronted with Elvov and to see my husband. I described the whole conduct of my case, from the first hearing by Party institutions to my imprisonment in the cellar, and ended by declaring my determination not to lie to the Party and not to ascribe to myself, let alone to other Communists, the fantastic crimes the interrogators had invented for a purpose unknown to me.

Long before I had finished, Major Yelshin was drooping with fatigue. He made a telephone call, and his relief arrived—it was Tsarevsky again. So it was to him I handed my statement.

He yelled, spat, swore, and reached for his revolver. But I knew they were not allowed to kill—at any rate, not until the end of the investigation. This I had learned from my dear instructress Lyama, who had explained it all to me in detail.

So I said nothing. Silently, I longed for my cell. But he kept me there until reveille at six.

Later, I was to learn what a lucky number I had drawn in the political lottery. My investigation was over by April, before the Veverses and Tsarevskys were authorized not only to curse and threaten their victims but to use physical torture.

15

Suddenly they stopped summoning me for interrogation. The empty prison days fell into a kind of regular routine, marked by the issue of hot water in the morning, the fifteen-minute walk in the prison yard (during which we were followed by guards with rifles and fixed bayonets), the meals, the washroom. The interrogators seemed to have forgotten my existence.

"They do it on purpose," said Lyama. "It's three weeks since I was last called. They hope prison life will drive you crazy, so that, in sheer desperation, you'll sign any old nonsense."

But I was so shaken by my first experience of Black Lake "justice" that I was glad of the unexpected respite.

"Well, let's make sure we don't go crazy," I said to Lyama. "Let's use our time to find out all we can about our surroundings. You said yourself that the great thing was to establish contact—and he's still tapping, isn't he?"

The prisoner in the cell to our left was still tapping on the wall, regularly, every day after dinner. But I had been too exhausted by the interrogations to listen properly to his knocking, and Lyama despaired of ever getting the hang of the prison alphabet.

One thing, however, we had noticed. On the days when our neighbor went to the washroom before us—this we could tell by the sound of the footsteps in the corridor—we always found the shelf sprinkled with tooth powder and the word "Greetings" traced in it with something very

fine like a pin, and as soon as we got back to our cell, a
brief message was tapped on the wall. After that, he immediately stopped. These knocks were altogether different
from the long sequences our neighbor tapped after dinner,
when he was trying to teach us the alphabet.

After two or three times it suddenly dawned on me:
"'Greetings'! That's what he's tapping," I told Lyama.
"He writes and taps the same word. Now we know how
we can work out the signs for the different letters." We
counted the knocks.

"That's right!" Lyama whispered excitedly. "The tapping comes in groups with long and short intervals. And
he tapped out nine letters in all: g-r-e-e-t-i-n-g-s."

During the long months and years I spent in various
prisons, I was able to observe the virtuosity that human
memory can develop when it is sharpened by loneliness
and complete isolation from outside impressions. One re-
members with amazing accuracy everything one has ever
read, even quite long ago, and can repeat whole pages of
books one had believed long forgotten. There is something
almost mysterious about this phenomenon. That day, at
any rate, after deciphering the message "Greetings" tapped
on the wall, I was astounded to find a page from Vera
Figner's memoirs * whole and fresh in my mind. It was
the page in which she gave the clue to the prison alphabet.
Clutching my head, myself astounded by my own words,
I recited as if talking in my sleep:

"The alphabet is divided into five rows of five letters
each. Each letter is represented by two sets of knocks, one

---

* Vera Figner, 1852–1942, famous woman revolutionary who took part
in the assassination of Alexander II and was imprisoned in the Schlüsselburg fortress in St. Petersburg for twenty years. The memoirs are about
her experiences during this time.

slow, the other quick. The former indicates the row, the latter the position of the letter in it."

Wild with excitement, interrupting each other and for once forgetting the guard in the corridor, we tapped out our first message. It was very short:

"W-h-o-a-r-e-y-o-u?"

Yes, it was right! Through the grim stone wall we could sense the joy of the man on the other side. At last we had understood! His endless patience had been rewarded. "Rat-tat, tat-tat-tat!" He tapped like a cheerful tune. From then on, we used these five knocks to mean "Message understood."

Now he was tapping his reply—no longer for a couple of idiots who had to have the word "greeting" repeated a hundred times, but for intelligent people to whom he could give his name:

"S-a-g-i-d-u-l-l-i-n."

"Sagidullin? Who's that?" The name meant nothing to Lyama, but it did to me. Much more boldly, I tapped:

"Himself?"

Yes, it was he—Garey Sagidullin, whose name for years past had not been mentioned in Kazan without an "ism" tacked on to it: Sagidullinism.

It was the heading of a propaganda theme. "Sagidullinism," like "Sultan-Galeyevism," stood for the heresy of Tartar "bourgeois nationalism." But he had been arrested in 1933. What on earth was he doing here now?

Through the wall my bewilderment evidently was sensed and understood. The message went on:

"I was and I remain a Leninist. I swear it by my seventh prison"—and startlingly: "Believe me, Genia."

How could he know my name? How could he, through the wall, in spite of all the strictness of our isolation, know

who was next door? We looked at one another in alarm.
We had no need to speak out loud. The thought was in
both our minds. He might be a *provocateur*.

Once again he understood and patiently explained. It
appeared that in his cell, too, there was a chink between
the window boards, and for a long time he had watched
us walking in the yard. Although we had never met, he
had once caught a glimpse of me at the Institute of Red
Professors in Moscow. He had been brought back to Kazan
for re-examination on additional charges. It looked like the
death sentence.

From then on, though outwardly nothing had altered,
our days were full of interest. All morning I looked forward to the after-dinner hour when the guards were
changed and, as they handed over their human cattle, were
for a while distracted from peeping through spy holes and
listening at doors.

Garey's brief messages opened a new world to me, a
world of camps, deportations, prisons, tragic twists of fate
—a world in which either the spirit was broken and degraded or true courage was born.

I learned from him that all those who had been arrested
in 1933 and '35 had now been sent back for "re-examination." Nothing new whatever had occurred to justify this,
but as the interrogators cynically put it, it was a matter
of "translating all those files into the language of '37"—
that is, replacing three- and five-year sentences by more
radical ones.

An even more important objective was to force these
"hardened" oppositionists (whose opposition in some cases
had consisted in advancing untested scientific ideas, such
as Vasily Slepkov's methodological research in the field of
natural science) to sign the monstrous lists, concocted by

the interrogators, of those whom they had "suborned." The signatures were obtained by threats, bullying, false accusations, and detention in punishment cells (the beatings began only in June or July, after the Tukhachevsky trial *).

Garey hated Stalin with a bitter passion, and when I asked him what he believed to be the cause of the current troubles, he replied tersely:

"Koba." (This was Stalin's Georgian nickname.) "It's his eighteenth Brumaire. Physical extermination of all the best people in the Party, who stand or might stand in the way of his definitely establishing his dictatorship."

For the first time in my life I was faced by the problem of having to think things out for myself—of analyzing circumstances independently and deciding my own line of conduct.

"It's not as if you were in the hands of the Gestapo." Major Yelshin's words rang in my mind.

How much easier and simpler if I had been! A Communist held by the Gestapo—I would have known exactly how to behave. But here? Here I had first to determine who these people were, who kept me imprisoned. Were they fascists in disguise? Or victims of some super-subtle provocation, some fantastic hoax? And how should a Communist behave "in prison in his own country," as the Major had put it?

All these anguished questions I put to Garey, ten years my senior in age and fifteen as a Party member. But his advice was not such that I could follow it, and it left me still more puzzled. To this day I cannot understand what

---

* Mikhail Tukhachevsky, 1893–1937, Soviet military leader who, together with Gamarnik, Yakir, Uborevich, and other members of the Soviet High Command, was executed in 1937.

could have prompted him, and Slepkov and many others who had been arrested in the early thirties, to act as he advised me to do now.

"Tell them straight out, you disagree with Stalin's line, and name as many others who disagree as you can. They can't arrest the whole Party, and by the time they have thousands of such cases on their hands, someone will think of calling an extraordinary Party congress and there will be a chance of overthrowing him. Believe me, he's as much hated in the Central Committee as here in prison. Of course it may mean the end for us, but it's the only way to save the Party."

No, this was something I could not do. Even though I felt obscurely, without having any proof, that Stalin was behind the nightmare events in our Party, I could not say that I disagreed with the Party line. I had honestly and fervently supported the policies of industrializing the country and collectivizing the land, and these were the basic points of the Party line.

Still less could I name others, knowing as I did that the very mention of a Communist within these walls would be enough to ruin him and orphan his children.

No; if the demagogic habits of mind I had been trained in were so deeply rooted in me that I could not now make an independent analysis of the situation in the country and the Party, then I would be guided simply by the voice of my conscience. I would speak only the truth about myself, I would sign no lies against myself or anyone else, and I would give no names. I must not be taken in by Jesuitical arguments which justified lies and fratricide. It was impossible that they could be of service to the Party I had so fervently believed in and to which I had resolved to dedicate my life.

All this—very briefly, of course—I transmitted to Garey. I had mastered the technique of tapping so thoroughly, by the end of a week, that Garey and I could recite whole poems to each other. We no longer needed to spell everything out. We had a special sign to show that we had understood, so we could use abbreviations and save time. A blow of the fist meant that a warder was about. I must confess that he used this signal much more often than I did, and I would surely have been caught if it hadn't been for him. However interesting the conversation, he never ceased to be on the alert.

I was never to set eyes on this man. He was eventually shot. I disagreed with many things he said and I never had a chance to discover exactly what his political views were. But I know one thing for certain: he endured his seventh prison, his isolation, and the prospect of being shot with unbroken courage. He was a strong man, a man in the true sense of the word.

16

• *"Can you forgive me?"*

I had been in prison for more than a month. After the first intensive spell of interrogation, I had not been summoned again except once, by a certain Krokhichev who handed me a note from my mother, saying only: "The children are well." He told me I had permission to receive parcels. Then, fixing me with his eyes—their lids red from night work like those of all the other interrogators—he mumbled something about a decision taken by the plenum of the Central Committee in February and March, which suggested that things mightn't be too bad after all, though, of course, one had to watch one's step.

My cheerful illusions were to be short-lived. The very next day Garey told us that our local masters had misunderstood the decision of the plenum. They had naïvely taken it in its literal sense, but new instructions had arrived and everything was to be understood in the opposite sense. More people were being arrested, and more brutal methods were to be used during interrogations.

One day the bolts and padlock on our door groaned open at an unusual time, after dinner, and two warders carried in an iron bed and set it up in the middle of the cell.

"They're moving a third one in," Lyama whispered excitedly.

"It's the busy season. They're short of space," I replied facetiously.

Ten minutes later the door rumbled again and a young woman came in, her cheeks spotted with red and her eyes wide with terror. Her face seemed familiar. She turned out to be Ira Yegereva, a postgraduate in hydrobiology, whom I had met once or twice at the university and knew as the spoiled only child of a professor. What could she have in common with political "criminals"? How on earth had she got here?

Four years earlier she had attended Slepkov's seminar and had even flirted a little with the clever, handsome professor. Now she was charged with belonging to a right-wing group. She was about as unpolitical as anyone can be, and couldn't have told right from left to save her life.

Hardly had she finished telling us her tragicomic story when Garey tapped a short message:

"I'm not alone."

His cellmate was Bari Abdullin, second secretary of the Party regional committee.

I had had a brush with him shortly before my expulsion from the Party, when I went to complain to the committee that my membership dues were being refused: the secretary of my Party organization was afraid of accepting them because I was under suspicion, although I pointed out that as long as I was not expelled I was bound to pay them.

At the regional committee I was received by Abdullin. I asked him what I should do: keep my card although my dues were refused, or turn it in and thereby invite fresh accusations?

Without looking up from his papers, he said in a tone which made further conversation impossible:

"The Party has every reason to distrust you, especially since you refuse to admit your mistakes."

Up to then we had been friends and had several times had adjoining villas in the summer. Now he was next door to me at Black Lake, sharing a cell with Sagidullin, whose name he had never uttered without pious horror.

The regional committee secretary, the pride of the Tartar working class! Could Garey be right in saying that Stalin had resolved to exterminate the whole Party elite?

That evening we learned from Garey's worried message that Abdullin was supposed to be guilty of pan-Turkism, contacts with the Turkish government, espionage, and presumably all kinds of other nonsense besides!

"They assume he's been plotting to cede the Kazan Province, as it was under the Tsar, to the Sublime Porte," Garey joked.

But we stopped joking when a few days later Garey tapped:

"Abdullin was on the conveyor belt for forty-eight hours" (the "conveyor belt" was continuous questioning by interrogators working in shifts), "and when he still refused to sign their balderdash, instead of bringing him back here they put him in the standing cell."

The "standing cell" was one of the "special methods" Tsarevsky had always been threatening me with. It was completely below ground where no ray of daylight would reach. In my simplicity I had thought it was called "standing" because it had no seats. In fact, it was a place so narrow that the prisoner could *only* stand upright with his hands at his sides.

"You mean he's walled in?"

"That's right."

We were so appalled that for two days we scarcely uttered a word. Even Ira stopped asking me what the "right-wing deviation," of which she was accused, meant. Nor did Garey knock on the wall. Our mood did not lift even when, as Krokhichev had promised, a parcel arrived for me. I stared at the bathrobe my mother had sent; it made me think of the seaside, of friendly, laughing people. Against this background I kept seeing the figure of a walled-in man—and not just any man but Bari Abdullin, who not long ago had been lecturing to Party activists on the international situation, who raced up and down in his garden with his little daughter on his back, who on Sundays played in the same volleyball team as I did.

At last we heard Garey tapping.

"They dragged him back unconscious. We were allowed to let the bunk down. They gave him a camphor injection. He's better now and asking for a cigarette. Have you got any?"

Yes, I had—two packs. I don't know what had given my mother the idea of putting them in the parcel, as I never smoked. Perhaps she thought I would have to in a place like this, or she may have had my cellmates in mind. Anyway, there they were. But how to get them to our neighbor?

Garey tapped out detailed instructions. If tomorrow we were taken to the washroom before they were, we must have the cigarettes with us, hidden under a towel. The one who carried them must go first, and the other two, walking in single file, must keep as far back as possible. When the first one came to the passage leading to the latrines and the shower room, she must quickly bend down and slip the cigarettes into the small space under the shower-room door. It was on the left. The one who came last must stumble over the threshold and hold up the guards.

We prepared feverishly for this delicate maneuver. First, we had an argument among ourselves. Ira, always cautious, was against trying to pass a whole pack. It would be noticed. It would stick out from the hiding place. We would all end up in the punishment cell. Lyama, on the contrary, was for going the whole hog. What was one pack to a man in his state? He should have both! And soap as well! Yes, I should give him the toilet soap Mother had sent. Let the poor wretch at least have a good wash after all those horrors. The men were sure to get even smaller pieces of soap from the guards than we did.

I took a middle position—either both packs without the

soap or the soap and only one pack. Otherwise we were sure to be caught. After a long argument we agreed on the soap and one pack.

"Then let's add a little butter. We've got over a quarter of a pound in the parcel, and you know how important food is to him now. Butter contains phosphorus. Good for the brain. It'll help him to hold out."

Dear Lyama! She had never done her basic course on Marxism—unlike Ira, who had recently taken her degree. Lyama was just a typist. And a lot of our prison leisure was taken up with her descriptions of the wonderful foreign clothes she had left behind. But whenever afterward I had to deal with scoundrels, it was the thought of Lyama, of her genuine fearlessness and generosity, that comforted me.

Wrapping our parcel was also a problem. The term "package" was purely conventional, as the guards had torn and thrown away the original wrapping. They handed us the cigarettes loose after making sure that none of them contained a note. The soap was also handed over unwrapped and stuck through with a penknife in several places. The butter was in a tin. Possession of the smallest scrap of paper was a serious crime.

How were we to tie the cigarettes together? We tried hair. Lyama and I pulled out a fair amount of each other's but, used as string, it slipped and came undone.

"What idiots we are!" Lyama struck her forehead. "We've got all the thread we need. There's my dressing gown." Good strong threads were pulled out of it, the cigarettes firmly bound together, and two thin slices of bread, thickly spread with butter, tied to the strawberry-scented soap.

The operation itself was brilliantly carried out. The

trickiest part—going first and slipping the parcel under the shower-room door—Lyama took on herself. I was to go last, keeping as far behind the others as possible and, above all, stumbling in the doorway to delay the guards. Ira, in the middle, had the task of walking as usual and not looking frightened.

It all went off superbly. Lyama darted in, while Ira, I, and the warder who followed us were still a long way behind, and had time to slip the package through the crack and even to check that it didn't show. I put on such a convincing act of stumbling and hurting my knee that the warder grunted: "Look out, can't you?"

The seven or eight minutes between our return from the washroom and that of our neighbors seemed endless. At last we heard the bolts and waited impatiently for them to be locked up. Finally the warder moved off down the passage.

"Rat-tat, tat-tat-tat" came Garey's joyful signal. Then slowly and distinctly: "You clever, brave, kind girls!"

We all three felt as soldiers must feel after a battle: tired and astonished at our own heroism. The first to recover from the sense of glory was Lyama, who began asking me whether Abdullin had a pretty wife and whether she wore nice clothes.

Toward evening the wall spoke in an unaccustomed voice. The tapping came in a slow, cautious, inexperienced hand.

"Genia! Genia!" said the wall.

It was Abdullin, clumsily tapping out a sentence which only I could understand:

"Can you forgive me?"

"Why is he asking that? Did you two have a love affair?"

While Abdullin, evidently shaken to the core by all he had been through, went on:

"How could you take such a risk? After all my callousness to you? What if you had been caught?"

"The thing is not to get caught. Learn prison techniques. Technology is the key to socialist reconstruction—remember?"

17

• The "conveyor belt"

They started on me again. I was put on the "conveyor belt"—uninterrupted questioning by a changing team of examiners. Seven days without sleep or food, without even returning to my cell. Relaxed and fresh, they passed before me as in a dream—Livanov, Tsarevsky, Krokhichev, Vevers, Yelshin and his assistant, Lieutenant Bikchentayev, a chubby, curly-haired, pink-cheeked young man who looked like a fattened turkey cock.

The object of the conveyor belt is to wear out the nerves, weaken the body, break resistance, and force the prisoner to sign whatever is required. The first day or two I still noticed the individual characteristics of the interrogators—Livanov, calm and bureaucratic as before, urging me to sign some monstrous piece of nonsense, as though it were no more than a perfectly normal, routine detail;

Tsarevsky and Vevers always shouting and threatening—
Vevers sniffing cocaine and giggling as well as shouting.

"Ha-ha-ha! What's become of our university beauty
now! You look at least forty. Aksyonov wouldn't recognize his sweetheart. And if you go on being stubborn we'll
turn you into a real grandma. You haven't been in the
rubber cell yet, have you? You haven't? Oh well, the best
is still to come."

Major Yelshin was invariably courteous and "humane."
He liked to talk about my children. He had heard I was a
good mother, yet it didn't look as if I cared what happened
to them. He asked me why I was so "becomingly pale"
and was "amazed" to hear that I had been questioned without food or sleep for four or five days on end:

"Is it really worth torturing yourself like that rather
than signing a purely formal, unimportant record? Come
on now, get it over and go to sleep. Right here, on this
sofa. I'll see you're not disturbed."

The "unimportant record" stated that, on Elvov's instructions, I had organized a Tartar writers' branch of the
terrorist group of which I was a member; there followed
a list of Tartar writers I had "recruited," starting with
Kavi Nadzhimi.*

"Anxious to spare Nadzhimi? He didn't spare you," said
the Major enigmatically.

"That's between him and his conscience."

"What are you, an Evangelist or something?"

"Just honest."

Again the Major couldn't miss the chance to display his
learning, and gave me a lecture on the Marxist-Leninist

* Tartar Communist writer, 1901–1957.

view of ethics. "Honest" meant useful to the proletariat and its state.

"It can't be useful to the proletarian state to wipe out the first generation of Tartar writers, who are all Communists at that."

"We know for a fact that these people are traitors."

"Then why do you need my evidence?"

"Just for the record."

"I can't put on record what I don't know."

"Don't you trust us?"

"How can I do that when you arrest me without cause, keep me in jail, and use illegal methods of interrogation?"

"What are we doing that's illegal?"

"You've kept me without sleep, drink, or food for several days to force me to give false evidence."

"Have your dinner, I'm not stopping you. They'll bring it this minute. Just sign here. You're only torturing yourself."

Lieutenant Bikchentayev, who now always accompanied the Major, evidently as a trainee, stood by "in readiness," repeating the final words of Yelshin's sentences like an infant learning to talk.

"It's all your own fault," said the Major.

"Your own fault," echoed the Lieutenant.

One day the Major prepared a questionnaire on my contacts with Tartar intellectuals.

"You're an educated woman, you speak French and German, why should you have wanted to learn Tartar?"

"So that I could do translations."

"But it's an uncivilized language. . . ."

"Is it? Is that what you think too, Lieutenant?"

The turkey cock smiled sheepishly and said nothing.

After this preamble I was invited to sign a statement to the effect that, on the orders of Trotskyist headquarters, I had tried to organize an opportunist alliance with bourgeois-nationalist elements of the Tartar intelligentsia.

I ventured on irony: "That's right! All my life it's been my dream to unite the Moslem world for the greater glory of Islam!"

The Major laughed, but he didn't give me anything to eat or drink, or let me go away to sleep.

It seemed to me then that my suffering was beyond measure. But in a few months' time I was to realize that my spell on the conveyor belt had been child's play compared with what was meted out to others, from June 1937 on. I was deprived of sleep and food but allowed to sit down and, occasionally, given a sip of water from the jug on the interrogator's desk. I was not beaten.

It is true that once Vevers nearly killed me, but that was under the influence of cocaine, when he was not responsible for his actions, and it gave him a great fright.

It happened, I think, on the fifth or sixth night. By then I was half delirious. As a means of "psychological pressure," the prisoner was customarily made to sit a long way off from the interrogator, sometimes at the opposite end of the room. On this occasion Vevers sat me down against the far wall and shouted his questions at me right across the vast office. He asked me in what year I had first met Professor Korbut, who had joined the Trotskyist opposition in 1927.

"I don't remember the exact year but it was a long time ago, before he voted for the opposition."

"Wha-at!" Excited by cocaine and my stubbornness, Vevers was beside himself with rage. "Opposition d'you call it, that gang of spies and murderers? Why, you . . ."

A large marble paperweight flew straight at me. Only when I saw the hole in the wall half an inch from where my temple had been did I realize what a narrow escape I had had.

Vevers was so frightened that he actually brought me a glass of water. His hands were shaking. Killing prisoners under interrogation was not yet allowed. His feelings had run away with him.

On the seventh day I was taken to the floor below, to a colonel whose name I cannot remember. Here, for the first time, I was made to stand throughout the interrogation. As I kept falling asleep on my feet, the guards placed on either side of me had continuously to shake me awake, saying: "You're not allowed to sleep."

A similar scene from a film called *Palace and Fortress* floated through my mind. The hero, Karakozov, was interrogated in exactly this way; he, too, was deprived of sleep. Then my head began to swim. As through a thick fog, I saw the Colonel's disgusted expression, and the revolver on his desk, evidently put there to frighten me. What particularly annoyed me, I remember, were the circles on the wallpaper, the same as in Vevers's office. They kept dancing before my eyes.

I have no idea what answers I gave the Colonel. I think I was silent most of the time, only repeating occasionally: "I won't sign." He alternated threats and persuasion, promising that I would see my husband, my children. Finally I blacked out.

I must have been unconscious for so long that they had to stop the "belt." When I awoke I was on my bunk and saw Lyama's dear, tear-stained face bending over me. She was feeding me drops of orange juice, which Ira had just received in a parcel.

Soon I heard anxious inquiries from Garey and Abdullin next door.

"Thank heaven she's come around. Splendid. Kiss her for us."

Supper was brought, and I ate two portions of the disgusting slop that passed for fish stew. By way of dessert, Ira triumphantly produced two small squares of chocolate from her parcel.

I had only just enough time to think how kind people were when I was called out for another spell on the conveyor belt.

18

• *Confrontations*

This time it was less severe and lasted only five days. For about three hours out of twenty-four I was allowed to return to my cell. True, this was never before six in the morning, so that when I got there the bunks were fixed to the wall and I could not lie down. But just sitting peacefully on a stool, with my head on Lyama's shoulder, eating a few lumps of sugar (the whole of our joint ration of six lumps was put at my disposal)—all this restored me a little, even though the warders carefully saw to it that I should never close my eyes. "It's forbidden to sleep by day," they explained.

On one of those days, Garey told us of the death of Ordzhonikidze.* I still don't know how he got his information, but I do know that when in 1956—after my rehabilitation and reinstatement in the Party—Khrushchev's report to the Twentieth Congress was read at a Party meeting, I heard the very same account of Ordzhonikidze's death as by wall telegraph from Garey in 1937.

My second spell on the conveyor belt was no more successful from the point of view of the interrogators than the first. I signed neither Yelshin's fantasy about the "reactionary cell of Tartar bourgeois nationalists" nor Vevers's about a terrorist plot against the secretary of the regional committee.

I do not want to sound like a heroine or a martyr. I am very far from thinking that my refusal to sign their lying records was due to any special courage on my part. Nor do I judge those comrades who, tortured beyond endurance, signed whatever was put before them.

I was simply lucky: the investigation of my case was over before the use of "special methods" became widespread. My obstinacy made no difference to my sentence: I got the same ten years as those who were tricked into signing the lists of people they had "suborned." But I did enjoy the blessing of a clear conscience and the knowledge that no one else was caught in the satanic web through my fault or weakness.

Having given up the attempt to extract a "sincere confession" from me, those in charge of my case left the task of rounding it off to Lieutenant Bikchentayev, who now called me out only by day. After two or three unproductive sessions, he told me importantly that, as I refused to

---

* Grigory Ordzhonikidze, 1886–1937, Georgian Bolshevik leader who committed suicide in February 1937.

confess, my evasions would be shown up in a series of confrontations which would begin next day. I was both intrigued and alarmed by the news—though as a rule it was difficult to be anything but amused at the turkey cock's announcements, so pompous was this comic-opera figure as he sat at his desk with its three telephones, his plump little face shining with foolishness that oozed from it like fat from a roast.

But to be confronted with witnesses! Could Elvov really be here? He might be, perhaps for a re-examination of his case. But what could he possibly say? It was one thing to sign lying statements about oneself, but how did one lie face to face with a comrade one had betrayed?

It was not Elvov, however, I found in Bikchentayev's office next day, but Volodya Dyakonov, a writer who had been on my staff in the cultural department of *Red Tartary*.

What on earth was he doing here? Had he been arrested? In any case, I was glad to see him. We were old friends. Our fathers had been schoolmates. I had helped him to get his job, and had gladly, almost lovingly, taught him his trade as a journalist. He was five years my junior. He had often said he was as fond of me as of a sister. It was good to see a face I knew so well, and even before Bikchentayev had dismissed my guard I held out both hands:

"Volodya! How are my children? Tell me quickly."

Bikchentayev rose from his chair, ready to explode at this unheard-of breach of the rules. A prisoner throwing herself into her accuser's arms! For Volodya, strange as it might seem, had been summoned to give evidence about my crimes. He was here to "confront" me.

"The following is order of confrontation," said Bikchentayev, whose Russian was not of the best. "I ask ques-

tion. Witness Dyakonov answers, then the accused . . ."
He pronounced my name in a guttural way and put the
stress on the wrong syllable.

"Volodya, so it's you I'm being confronted with! What
can you possibly say against me? Or have they arrested
you too, and forced you to sign a pack of lies accusing us
both?"

Bikchentayev thumped the table with his fat, dimpled
fist, but it was funny, not frightening.

"Accused! Stop exerting pressure on witness. And you,
Dyakonov, behave as told, or you get arrested and go to
jail."

So Volodya was not under arrest. Then what was the
point of this farce? But I could see from his expression that
it was not a farce. His face was livid, his eyelids twitched,
his lips were blue and trembled.

Instead of replying to my question about the children,
he stammered:

"I—I—I'm sick, Genia. I've just had encephalitis."

"Witness Dyakonov," Bikchentayev solemnly opened
the proceedings. "Yesterday at interrogation you stated
that among editorial staff of newspaper *Red Tartary* there
was subversive counter-revolutionary terrorist group, and
accused was member. Do you confirm this now in presence of accused?"

It was frightening to watch Volodya's face, its handsome features so contorted by a nervous tic that it looked
almost hideous. He mumbled:

"What . . . what . . . I actually said was that the people you were asking me about held important jobs on the
paper. That's all I know."

Grimly knitting what would have been his eyebrows if
he had had any, Bikchentayev turned to me:

"You confirm this?"

"What is there to confirm? He simply told you who were the important people on the staff. The only person who said anything about terror and subversion is yourself. The witness hasn't breathed a word of it."

Bikchentayev grinned maliciously and wrote down, first his question, then Volodya's answer, worded as follows: "Yes, I confirm that there was a subversive counter-revolutionary group among the editorial staff of *Red Tartary*." He pushed the paper to Volodya.

"In a confrontation, each question and answer must be signed separately. Sign!"

Volodya hesitated, his hand shaking so that he could scarcely hold the pen.

"Volodya," I said mildly, "you know it's a trick. You never said anything of the kind. By signing this you'll be causing the death of hundreds of your comrades, people who have always been decent to you."

Bikchentayev's eyes nearly popped out of his head.

"How dare you exert pressure on witness! I send you straight away to the lowest punishment cell. And you, Dyakonov, you signed all this yesterday when you were alone here. Now you refuse! I have you arrested at once for giving false evidence."

He made a show of reaching for the bell—and Volodya, looking like a rabbit in front of a boa constrictor, slowly wrote his name in a hand as shaky as though he had had a stroke and quite unlike the bold sweep of the pen with which he signed his articles on the moral code of the new age. Then he whispered almost inaudibly:

"Forgive me, Genia. We've just had a daughter. I have to stay alive."

"And what about my three children, Volodya? And the children of all those people on your list?"

Bikchentayev was again ranting and thumping his desk, but I was not afraid of him. Comic roly-polies should not be appointed to these flesh-creeping jobs, they don't carry conviction. I went on:

"And what about yourself, Volodya? If you'd really known of such a group and not reported it till you were summoned—from 1934 to '37—then you would have been abetting it, and that's a crime."

Volodya's face grew paler and bluer. Tears rolled down his cheeks and he did nothing to hide them. As for Bikchentayev, whose patience was now exhausted, he rang for the guard and ordered him to take me to a punishment cell, but at this moment Tsarevsky came in and whispered something in his ear.

I was taken out into the passage, and when they brought me back after five minutes Volodya had gone, and in his place . . .

Well, this certainly was a day of surprises. In Volodya's place I found a very old friend of mine, Nalya Kozlova. She too had been on my staff at *Red Tartary* and had also got her job through my backing. As students we had been inseparable. In those days she was always writing something and had been nicknamed the Author. How many exams we had prepared for together, how many poems we had read to each other, how many confidences and reflections on life we had exchanged! And now here she was, together with Volodya, to give my persecutors a hand!

I felt I was choking. Was this a conspiracy of all the demons in hell to turn me from a thirty-year-old woman into an old crone of a hundred and make me say, with

Herzen: "Everything is destroyed, the freedom of the world and my happiness"? But perhaps Nalya had thought of some clever way of rescuing me, some way I did not yet understand? I tried eagerly to catch her eye, but she looked away.

This time, Bikchentayev had nothing to complain of.

The witness was no weak-willed, tearful Dyakonov. Used to newspaper work, she was so articulate and fluent that it was all he could do to keep pace with her. Before he could draw breath she was signing a statement confirming that there had been a subversive terrorist group at work in the office, and that I had been an active member. She even added that Kuznetsov, the editorial secretary, had been the organizer, while I was in charge of political indoctrination!

Smiling craftily, Bikchentayev put the question intended to deal me the *coup de grâce*:

"Do you consider that the defendant's subversive contacts were formed accidentally, or did she already have them in her student days?

And my friend Nalya—my dear, witty, dashing fellow student—recited like a parrot:

"No, her links with the Trotskyist underground cannot be considered accidental. Among her friends in her early youth were Mikhail Korbut and Grigory Voloshin, both since proved to be political criminals. They were no doubt mainly brought together by the similarity of their political views."

Suddenly, all three telephones on Bikchentayev's desk rang. Drunk with the glory of his historic role, our young Robespierre put a receiver to each ear and shouted "Wait!" into the third.

I seized the opportunity and said softly to Nalya in French (both of us had taken it for our honors degree):

"You've certainly chosen a noble part! Good enough for a film or a novel by Dumas père! Have you gone clean off your head?"

Also in French, Nalya answered without raising her eyes:

"If you don't stop insulting me, I'll talk about Grisha Berdnikov as well." Grisha Berdnikov had been a Party member since 1917. In recent years he had worked in Sverdlovsk. The fact that Kozlova was using his name to threaten me probably meant that he was under arrest. She must have thought the connection particularly damaging because Grisha worked on *Izvestia* under Bukharin. I had known him neither better nor less well than had anyone else in our office, but she realized that the mere mention of still another "political criminal" would damage my case. Suddenly I lost my temper.

"Just you dare," I hissed. "If you do, I'll change my tune and sign any balderdash they cook up, and I'll name you as an active member of the group."

At this point Bikchentayev, having torn himself away from his telephones, caught the sound of a foreign language being spoken in his office.

"In what language you bring pressure on the witness?"

"In French."

More thumps of the plump fist, more ravings about punishment cells . . .

"I'm so sorry, Lieutenant," I said sweetly. "I was only quoting a proverb—something like 'You live and you learn.' I had no idea you didn't understand French."

The witness Kozlova cast a frightened glance at me.

How could I make fun of a man who held my fate in the palm of his hand! But I knew I was taking no risk. Bikchentayev was too stupid not to take my remark at its face value, as indeed he did.

"Nobody say nobody not understand," he said, mollified. "But official language of interrogation is Russian, kindly stick to it. You have proverb, you say it in Russian."

He remained in a good mood till the end of the interview, when he handed the complete record to Kozlova to sign at the bottom of the last page. I watched the faint splutter of ink over the signature I knew so well. He blotted it neatly and, with a gracious smile, handed Kozlova her pass:

"You free to go, comrade."

Near the door she suddenly hesitated and, blushing, held out to me her rolled-up newspaper:

"Would you like it? It's today's."

"No, thank you. We aren't allowed to read newspapers in prison. Nor books."

The telephone shrilled again so Bikchentayev had no time to deal with me. He picked up the receiver and pressed the buzzer for the guard.

Nalya still hung about. The detail she had just learned (no reading!) had evidently opened her eyes to something that had never occurred to her.

"So you've heard no news at all?" she whispered quickly while Bikchentayev was still on the telephone. "Ordzhonikidze's dead. So is Ilf."

"They're lucky. They've just died. I'll be shot now, thanks to your and Volodya's lies about me. . . ."

Nalya backed to the door, her eyes filled with horror.

Only the turkey cock would have let me get away with a conversation of this sort. Vevers or Tsarevsky would have sent me to the punishment cell for a week at the very first attempt. But Bikchentayev only gave a squeal and, no longer the well-bred gentleman, told Kozlova to be off quick, before he canceled her pass. He even forgot to send me to the punishment cell, so pleased was he with the outcome of the interview.

I returned to the cell so shaken that I could not answer Lyama's questions or Garey's knocking. Night fell. There is nothing more frightening than prison insomnia—I had been suffering from it ever since my spell on the conveyor belt.

I listened to the even breathing of my cellmates and the measured creak of the warder's boots along the corridor. From time to time a padlock ground open, and there were footsteps and whispers. Someone was being taken up for interrogation. Every sound echoed in my temples.

Nalya! What fun we had had, racing each other as we swam in the country, or getting into concert halls without tickets! We were eighteen then, and good friends.

Dawn came at last. The sun struggled through the bars and the wooden screen, and the minutest glimmer showed on the dirty gray walls, looking like a tiny gold beetle on a big manure heap.

By now it was April. Spring. The spring of 1937.

19

• *Parting*

The day began as usual—inspection, washroom, hot water, bread. It was, in fact, better than usual, for Lyama had brought off the *coup* of stealing a needle from the head warder. We were allowed the use of a needle for only five minutes once a week. The "chief" who issued them always had several stuck into the outer pocket of his tunic. He came around each morning to count heads and inspect the simple furniture of the cell. He pulled out the night-table drawer, lifted the straw pillows, and even looked into the slop pail.

So he did that day. But while he was bending over the drawer, Lyama—with a snakelike movement of lightning speed—managed to pull one of the needles out. We had threaded it (with thread from my bathrobe) and were quietly mending our stockings when, at this untimely hour, the door opened.

"Bring your things along."

I was being moved out. . . . "Bring your things" meant it was for good. Indescribable excitement gripped all of us.

"They're releasing you! You're going home!" cried Ira, always the one most given to illusions. "Go and see my people. And to show that you've been to see them and they know about me, tell them to put a 'snowflake' candy in my parcel." Lyama, who had gone pale, snapped at her:

"Don't be silly. Is it likely, after all those confrontations? How do we know it's not the punishment cell? Or deportation?"

As usual, it was Garey who knew:

"There's a Black Maria in the yard. It's taking people whose interrogation is over to the Krasin Street prison, to make room for new arrivals."

That day I first experienced the pain of prison partings. There are no more fervent friendships than those made in prison. Now these ties—almost closer than those of flesh and blood—were being severed. The same pitiless hands which had taken away my children, my husband, my mother, were robbing me of my sister Lyama and my faithful friend Garey. We were parting for ever, losing all trace of each other, as if going to our death—and quite likely we were. Except perhaps for Lyama, our chances of escaping capital punishment were not great.

"A scarf to remember me by, Genia darling."

Her hands trembling, Lyama pushed a square of Chinese silk into my pocket. I gave her my scarf in exchange. For a brief moment we embraced, sobbing.

The scarf, a tempting article of foreign manufacture, was later stolen from me by a pickpocket in camp. Lyama I never saw again, nor do I know what happened to her. But I shall never forget her golden hair, her deft, kind hands, or her big, glowing brown eyes.

Garey (he was alone again—Abdullin was being tortured on the latest form of "conveyor belt") communicated his feelings through the thick stone wall. Words of friendship and devotion came to life on it and, as usual with him, they sounded rather grandiloquent.

"Good-by, dearest. I wish you courage and pride. I believe that the ties of prison kinship are indestructible. I'll remember you until death. That may not be far off. Yet who knows . . . fetters may fall and prisons crumble. . . ."

There was a great commotion in the corridor where the transfer was being organized—doors banging, bolts creaking, guards whispering. Under cover of all this noise it was easy to tap out a farewell message to Garey.

The cell door closed behind me. Surprisingly, I was given back my watch. It had not been wound since the fifteenth of February and still showed two o'clock on that memorable day—the day on which my life ended. Everything since then had been only my wandering, after death, through hell. Or could it be purgatory? Could Garey be right about fetters falling and prisons crumbling someday? What would have become of us all if it had not been for the illusory light of that tenacious hope?

20

• *New encounters*

So this was the Black Maria—a navy-blue truck for the transport of prisoners. I had often seen it in the street and paid no attention. I thought it carried milk or groceries.

Inside, the truck was divided into two rows of tiny, pitch-black, airless cages. A prisoner was stuffed into each. Our bundles were piled up in the narrow passage between the two rows.

Kenneled as I was, I still remembered Garey's lessons. At once, without allowing myself to dwell on the horrors

of our situation, I set about establishing contacts. While
the warders' boots were still stamping outside, I tapped
on the partitions to my left and right: "Who?" "Who?"
From the left came the reply:

"Yefrem Medvedyev."

What luck! Yefrem was a postgraduate student I had
known at the Marxist Institute.

"When?"

"April 20th."

Quite recent! Now I could find out what had been
happening in town since my arrest.

We soon found that we did not even have to knock but
could simply whisper. We could hear each other perfectly,
and the noise of the engine prevented the warder, who sat
in the passage, from hearing us. I could hear the real voice,
warm and vivid, of Yefrem.

"Hello, Genia. I met your husband in the street at the
beginning of April. He'd been to Moscow to try to get
you out, but it was no good. Your children are well, but
the older ones miss you a lot."

"Who's been arrested since I was?"

"You might as well ask who hasn't!"

He rattled off dozens of names of Party activists, engineers, teachers.

Beyond the partition to my right someone was complaining in Tartar. For a long time he wouldn't answer my
questions, but finally he plucked up courage and told me
his name. I didn't know him. He was the district committee chairman of some rural area.

The drive took a fairly long time. I felt oppressed and
half stifled, but Yefrem's voice distracted me from my
discomfort.

"Yagoda's in jail too," he said. "Yezhov has replaced

him—the one who was chief organizer of the Executive Committee. He's turned out to be a real bastard, by all accounts—they say it's he who is responsible for all this."

The Black Maria stopped. One by one we were led out, to be swallowed up by the black gateway of the ancient prison which, in its day, had housed the rebels led by Pugachev.*

The same routine as at Black Lake—form filled in, watch removed (I need not have bothered to wind it). By an oversight on the part of the guard, we had a "collision"— an unscheduled meeting between prisoners. I caught sight of Professor Aksentyev, the head of the Tuberculosis Institute, his chin covered with black stubble. We had no chance to talk. The frightened warders literally dragged us off in opposite directions.

Every monastery has its own rules. In this one they took away not only my watch but my garter belt as well. A nurse with a caseful of medicines, whose additional function was to search women prisoners, wrinkled up her freckled little nose in compassion:

"It's so different from the old days. We used to get thieves, prostitutes. Now it's such educated ladies, you feel really sorry for them. Here's a bit of bandage to hold up your stockings—how they expect you to walk about without garters . . . only don't let anybody see."

She stole a glance over her shoulder, to make sure we were alone in the tiny dispensary where the women were searched:

"What made you do it, for goodness' sake? Why did you rebel against the government? I know you're Aksyonova, the wife of the town committee chairman. What did

* Yemelyan Pugachev, 1742–1775, leader of the Cossack rising in the time of Catherine the Great.

you want that you didn't have? Cars, villas, foreign clothes . . ."

Evidently she couldn't think of any more luxuries. I smiled warily.

"It's all a misunderstanding. The investigators made a mistake."

"Hush!" she glanced at the door. "Is it true what my father says, that you stood up for poor people, for peasants—so as to get something done for them?"

Fortunately, the arrival of the wardress saved me from having to answer. But it was interesting that people were trying to find some sort of rational explanation for what was going on.

The wardress took me up a worn stone staircase to the second floor. Although we were aboveground, the smells of dirt, mold, excrement, were even stronger than in the cellars of Black Lake. These and some other indefinable smells together made up the unique, indescribable stench of jail.

If the dirt and stench were worse than at Black Lake, the discipline was less strict; you could feel it at once. The prison had long been used for non-political prisoners and had not yet adapted itself to the new requirements. Imagine a conversation with the silent warders at Black Lake, such as I had just had during the search!

Fairly loud voices came from the cells. The warder who received me on the second floor didn't look as if he were made of stone. He looked me over with a mixture of cheerful curiosity and sympathy, and said kindly:

"Let's try Number 6, the women are a bit cleaner."

This too could never have happened at Black Lake.

Later on I was able to establish a general principle: the dirtier the prison, the worse the food, the ruder and more undisciplined the guards, the less danger there was to life. The cleaner the jail, the more we got to eat, the more courteous the jailers, the closer we were to death.

Here the doors were made of wood, not iron; they had big dusty spy holes, and the padlocks were not as huge as at Black Lake.

"Here's a new guest for you!" announced the warder, actually smiling.

The door locked behind me. I looked around. What a lot of them there were! I was immediately bombarded with questions. From a corner came an oddly triumphant voice:

"Welcome, welcome—it's Aksyonov's wife!"

A thin, bent, gray-haired woman, plainly delighted that I was in jail, got up, a cigarette in her mouth, and held out her hand.

"Derkovskaya, member of the party of Social Revolutionaries," * she introduced herself. "I've met your husband. I had to go to his office once. Little did he think that his wife would be sharing a cell with me in a few months! Well, frankly, I'm glad that some of you Communists are in jail at last. Perhaps you will learn in practice what you couldn't understand in theory. But you'd better get settled now. We'll talk later."

Getting settled was not so simple. The cell, which was meant for three, already held five; I was the sixth. In addition to the three wooden bunks along the walls, a single large plank bed had been set up in the middle. As my cellmates were shifting their rags to make room for me, the door creaked again and Ira Yegereva was brought in. The Black Maria had made a second journey and brought

---

* Social Revolutionaries: populist socialists who were regarded as bitter enemies by the Social Democrats, Bolsheviks, and Mensheviks. The party was suppressed in 1918.

another batch from the cellars at Black Lake—people whose investigation was over or on the point of completion.

Ira's arrival at once distracted attention from me. She was still well dressed. Tsarevsky—whether out of secret admiration for this delicately nurtured girl, a professor's daughter, a being from another world, or out of gratitude for the political innocence which made her swallow his simple arguments and sign whatever he wished—had allowed her to have weekly parcels.

In the course of settling in and unpacking, Ira showed her dresses to her new companions, and told them the story of each. Sweet memories floated on the stinking air:

"This one I used to wear to play tennis in at Sochi last year. Afterward it got a bit tight for me, but now it's just right. I've been losing weight."

The one who proved most responsive to Ira's reminiscence of seaside and tennis was a tall, plumpish young woman whose round face made me think of Maupassant's *Boule de Suif*. This was Anna, known in the cell as Big Anna, to distinguish her from Little Anna, whose bunk stood against the opposite wall.

Big Anna came from Moscow and had worked as a clerk at the railway office in Kazan. She was twenty-eight. She had neither husband nor children, only a certain Vova from whose bed she had been dragged at dawn a month before, by the investigator who arrested her. Vova turned white as a sheet: "What have you been doing?"—"Absolutely nothing," said Anna airily and, planting a kiss on Vova's trembling lips, she went off fearlessly with the investigator. They had brought her over in a car.

" 'What are you charging me with?' I asked him. 'Is it my morals or my politics?'—'Politics,' he says. 'Well, if it's that,' I said, 'you've come to the wrong address, you'll have to apologize.' The bastard just grunted, and do you know what it was? Just a couple of anecdotes I'd told. Seven years I'm getting, three and a half for each."

She then told us the anecdotes. The interrogator had drawn up two separate charges—one for *lèse majesté* (against Stalin), the other for slandering the collective farm system. Anna lost her temper and told him to his face: "So I did make a joke, so what? It wasn't at a public meeting, it was at home at dinner with my friends. And anyway, are you pretending that it's not true? I can't see that you're so anxious to get into a collective farm." So saying, she had signed both charges and was now awaiting her trial and the promised seven years' sentence.

Big Anna was the first representative I met of that great army of "babblers"—people who told political jokes and fell under the "lenient" provisions of Article 58(10) of the Penal Code; not being Party members, they enjoyed an advantage over us "terrorists," "spies," "diversionists," and "saboteurs." In our everyday prison life she was the friendliest of creatures, tolerant and easygoing, with a cheerful if somewhat cynical sense of humor.

When her neighbors felt too wretched to talk to her, she never got annoyed. She sang to them instead. Her favorite songs were "Bananas and Lemons in Singapore" and a sad one called "The Summerhouse." When her voice, turning from a natural soprano to a rich contralto, boomed "you'll come no mo-o-o-re," her nearest neighbor, Lydia, groaned as though she had toothache.

Lydia Georgievna Mentzinger was fifty-seven and in prison for the third time. She came of a German family long settled in Russia, had taught German, and was a fanatical Seventh-Day Adventist. I can still see her huge eyes filled with the deepest despair. They made me think of Andreyev's * story about the resurrection of Lazarus: it describes Lazarus's friends and relatives, all rejoicing at the miracle, while he sits among these cheerful people, looking at them with the same expression as Lydia's, because he has experienced death and knows what it is.

Somewhere in these pages I have said that my intense curiosity about life in all its manifestations—even in its debasement, cruelty, and madness—sometimes made me forget my troubles. I noticed the same thing about many of my companions along the cruel road. Many of them, too, still kept their illusions. What was happening was too absurd to go on for long, they believed. And this expectation that the colossal misunderstanding would suddenly be dispelled, the gates suddenly be flung open, and each of us rush back to her own hearth, cold though it was, kept up their courage.

But Lydia had no curiosity and no illusions. She knew perfectly well that there was nothing to hope for, and she knew that nothing would happen to excite her curiosity—she had seen it all before.

Later, I was to meet a great many believers of various denominations. All of them tried to propagate their faith and make converts. Lydia alone did not. She sat silently for days on end, with her feet tucked up on the bunk, staring over our heads with her Lazarus's eyes.

Little Anna was a Party activist who specialized in women's welfare work.

"I've always been in the Party," she told us, brushing back her straight brown hair, cropped in the Party-activist

---

* Leonid Andreyev, 1871–1919, novelist and playwright.

style. "First I was a 'Young Octobrist,' * then in the Komsomol, then in the Party."

And indeed it was impossible to imagine her outside the Party, outside the specific style and way of life which had developed in Party circles in the twenties and thirties. Now she was always forgetting where she was. Either she got carried away, telling us how successfully she had reorganized the work of women at a textile mill, and made plans for further improvements, deciding which of the women to recommend for promotion, or she was regretting that she had not transferred to a rural district where there was much more scope for her work, as the secretary of the rural district committee had urged her to. The secretary was now in jail, in the cell immediately below ours, on the first floor.

On her return from the interrogations Little Anna would lie silently on her bunk, gray-lipped, with her face to the wall, till nighttime. In the middle of the night she would come over to me, lie beside me, and whisper excitedly:

"Sh-sh-sh, Genia dear, I don't want the non-Communists to hear me, they might get the wrong idea . . . they're a terribly mixed lot here, you know. Some are Social Revolutionaries! But just listen . . ."

She was accused of "sabotaging Party work" and of "association with an enemy of the people." The "enemy" was the secretary of one of the Kazan municipal committees, who happened also to be her husband—the handsome and adored husband of simple Little Anna, who wasn't even pretty.

"I never did understand how Vanya ever fell in love with me—all the girls were after him. But we've lived to-

* Communist organization for very young children in the first grades at school.

gether for seven years and he still loves me. I can see he does. He loves me for my soul, for my Communist heart. And now the interrogator . . ."

She choked with sobs. The interrogator was saying that her marriage had been fishy from the start. A handsome fellow like that and plain Little Anna! It must have been a bogus marriage contracted on orders from the anti-Soviet underground.

"And what about my little Boris and Lydia? Are they bogus too?"

I stroked her shoulder, as skinny as a child's.

"Don't you listen to that scoundrel. All the Party activists know how much Vanya loves you."

"Sh-sh-sh. Please don't say things like that. Nina might hear you—that working girl who's never been in the Party. If we Communists say such things about the interrogators, what's she going to think?"

But Nina Yeremenko slept like a log, only occasionally crying out in fright. On the other hand, in the daytime she got on our nerves by swaying backward and forward on her bunk, clutching her knees and saying over and over: "When will this end?"

None of the theoretical arguments which distracted us interested Nina. None of Ira's reminiscences woke a response in her. Black Sea beaches and tennis were too remote from the leather-goods factory world of this raw-boned girl with the irremovable stains on her hands and the ineradicable (though by now two-months-old) smell of rawhide she exuded.

Of her twenty years, Nina had spent five at the Spartacus factory. What had ruined her was a birthday party to which the red-haired Lelka had invited her; like a fool, she had gone. Well, actually, it was on account of Mitka

Bokov. She'd been out with him a few times, and he wasn't the flighty sort, he was always talking about establishing a family. "I'd never let my wife go out to work," he said. "She'll just be a housewife." So she went to this party, knowing he'd be there, and she even got Lelka a brooch from the jeweler's—a really good silver gilt one. Well, the boys had a bit to drink, and apparently one of them said something about Stalin. So help her, she never even heard what it was. And now they'd come down on her with this Article 12 for "non-denunciation." "Your duty as an honest Soviet proletarian," they said, "was to go next day to report it to the NKVD, instead of which you hushed it up."

So for the past two months Nina had sat clutching her knees and repeating: "When will this end?" Nothing cheered her up. She wouldn't even accept the candies Ira offered her from her parcel, and if Big Anna sang too poignantly about the summerhouse, Nina burst into tears. What terrified her most of all was that Mitka might get tired of waiting for her and marry someone else, and Nina's dream of living happily as a housewife who did not go out to work would be destroyed for ever.

We tried to comfort her by pointing out that, very likely, Mitka was himself in prison: after all, he too had failed to denounce whoever it was. At this her sulky face lit up from within, glowed like a pearl, and looked almost beautiful as she tried passionately hard to convince us that Mitka would never be arrested—they couldn't possibly do without him at the factory. God forbid that such a thing should happen! She'd even sooner he married Lelka. So long as he was safe and sound . . . Let Nina alone suffer, if that was to be her fate.

But what terrified Little Anna—who was used to working with just such girls as Nina—was that Nina might develop "an unhealthy attitude" toward the Party as a whole. This was why Anna confided her troubles only to me, "as one Party member to another." But the one thing that frightened her even more was the idea of being overheard by Derkovskaya, the Social Revolutionary.

"You realize, Genia, she really is *the* class enemy. Mensheviks and SR's, you remember. Though of course, from what the schoolbooks said, I didn't imagine them quite like that. She's really a good old soul, and dreadfully unhappy. But one mustn't let one's mind be swayed by pity . . . above all, we mustn't play into the hands of enemy propagandists."

I certainly was swayed by pity, especially when it came to the story of Nadezhda Derkovskaya's son Volodya, born in a Tsarist prison where his mother was in solitary confinement in 1915. Except for short intervals, both his parents—Social Revolutionaries—had been in jail since 1907. The family was set free by the revolution of February 1917 when, for the first time, Volodya saw his mother's native city, Petrograd. But by 1921 they had once again been deported, and Volodya's father died in Solovki.* Wandering from one place of exile to another, Volodya and his mother came to Kazan. There he had his last lucky break—until 1937. For the nth time Nadezhda was arrested, and Volodya—twenty-two and, to his mother's joy, newly admitted into the teachers' training college—was arrested with her.

"His only crime is that he was born in a Tsarist prison and grew up in exile," said Nadezhda. "He's not a Social

---

* Island in the White Sea, which in 1917 became a notorious place of detention for political prisoners.

Revolutionary, he's never been interested in politics, he's just a brilliant mathematician. The reason he has followed me about is that he's fond of me. After all, we've got no one else."

With dreadful clarity, I imagined all this happening to my own Alyosha. It was unbearable. It is still just possible to live, inwardly resisting, when you yourself have been snatched and swept away by some evil power which seeks to rob you of your health and reason, and either turn you into a dumb beast of burden or kill you. But when this happens to the child you have reared and cherished . . . So I felt a devouring pity for Derkovskaya, the first Social Revolutionary I had ever met—although she made no bones about telling me exactly what she thought.

"I liked your husband better than any of the other Communist officials that I've seen," she told me, lighting one cigarette from another. "I went to see him when I lost my job, and he talked to me like a human being. I feel sorry for you personally. But I tell you frankly that I'm glad the Communists are beginning to get a taste of their own medicine."

I was curious to know what alternative the Social Revolutionaries were putting forward as an opposition nowadays. But I soon realized that they had no positive program of their own. All Derkovskaya had to offer was criticisms of the existing regime. What held the Social Revolutionaries together were the links they had forged in countless prisons and places of exile. Afterward, in camp, I had many opportunities to convince myself of the strength of these bonds, which made them almost a separate caste.

One day Derkovskaya ran out of cigarettes. Used to chain smoking, she was in torment. Just then I got a parcel from my mother in which she had put the usual two packs.

"Saved!" I said cheerfully, holding them out. Derkovskaya blushed and with a muttered "Thank you" turned away:

"Just a second. I won't be long."

She sat down by the wall and tapped a message. One of the prisoners in the next cell was Mukhina, the secretary of the Social Revolutionaries' clandestine regional committee. Derkovskaya tapped away, not realizing that I could follow:

"There's a woman Communist here who has offered me cigarettes. Should I accept?"

Mukhina inquired whether the Communist belonged to the opposition. Derkovskaya asked me, passed on my reply —and Mukhina tapped categorically: "No."

The cigarettes lay on the table between us. During the night I heard Derkovskaya sighing deeply. Though thin as a rake, she would much sooner have done without bread. As I lay awake on my plank bed, the most unorthodox thoughts passed through my mind—about how thin the line is between high principles and blinkered intolerance, and also how relative are all human systems and ideologies and how absolute the tortures which human beings inflict on one another.

21

• *Orphans twice over*

As I have already said, it was for the first time in the twenty years since the revolution that the prison in Krasin Street was being used for political prisoners. Until 1937 there had been plenty of room for them at Black Lake. Now, all three prisons in Kazan were packed with "enemies of the people." But the old traditions of a non-political prison—dirt, rudeness, and a certain laxity of discipline—survived by inertia.

Here, you could knock on the walls to your heart's content—the light sound of tapping was drowned by the general racket in our hot, stinking, overcrowded hell (at Black Lake even Garey's pinpoint signals seemed to echo through the cell). If the guards scolded us, it was only halfheartedly.

As a result, we were soon in touch with almost the whole prison. The panes of the ancient window were broken, and the wooden screen was of a different shape from the one at Black Lake. It was much wider at the top, thus letting in more light as well as acting as a sounding board. If you stood right up against the window and spoke straight at the screen, you could be heard clearly in the cell below.

But to speak as openly as this was too dangerous. So we invented a so-called "operatic" method of communication. The inventor was the rural committee secretary, Sasha—I forget his surname—who was an inmate of the cell immediately below ours.

Once, at the end of a stifling hot day, while the warders
were busy dishing out the soup, we heard a passable baritone singing the Toreador's aria from *Carmen* in the following unusual version:

How many are you, pris'ners up there?
How many there? Please to declare!
Te-e-ell us who and what you are,
Tell us, tell us, we implore!
Our hearts are all aglow.
Your names, yes, yes, your names we want to know.

The message was not hard to understand, and before
long first our names, then theirs, were sung to a variety
of tunes. Once this vocal link had been established, we
kept closely in touch and had soon learned all the news,
of which there was plenty. Every day we heard the names
of new arrivals, what they were charged with, and how
quickly the "special methods" were spreading. We managed also to organize an interchange of notes through the
washroom. We wrote them on tiny scraps of cigarette
paper, still using a pencil stub Lyama had once stolen from
an interrogator and given me before I left.

At first Sasha was full of Olympian self-assurance and
looked on the current troubles as a slight misunderstanding
which would soon be cleared up. In his duets with Little
Anna he even continued to urge her to join "his" committee, as soon as "we are both out of here," listing, in his
magnificent deep voice, the advantages of the job he
offered over the one she had held before. Speaking of two
of his cellmates, engineers who were not Party members,
he could not rid himself of a patronizing tone—even
though the three of them shared the same bunk and took
turns carrying the bucket of excrement out to the latrine.

I do not mean that Sasha was a fool. His attitude only went to show the force of inertia and of the hypnotic power of impressions received early in life.

His awakening, like that of thousands of other Sashas, came when "active methods" of interrogation were applied to him. One day one of the non-Party engineers told us, to a tune from *Prince Igor*, that Sasha had come back with a split lip. It was swollen and bleeding. Had we, perhaps, any ointment, such as vaseline, and could we spare a few cigarettes?

We did have cigarettes, but how were we to get them to him? The washroom was a foul, stinking hole—we couldn't leave them there for Sasha afterward to put into his mouth. We hit on the idea of letting them down on a thread through the window. My bathrobe lost a few of its remaining threads, and we dangled the cigarettes one by one, like bait at the end of a line, through the crack at the bottom of the screen. Sasha's friends deftly caught two of them with a wooden spoon. But the third got stuck between the floors, and when we went out for recreation we could see it catching the sun. As soon as we got back to our cell, we improvised:

Up above your window, Sasha,
Shines a cigarette so new.
Quickly go retrieve it, or
We'll all be in the stew.

To our relief, a rich baritone replied:

I hear and understand.
The fair one
Will be rescued before dawn.

We giggled like schoolchildren.

Yet it was at such a moment, when we had all been laughing as though our troubles were over, that I was struck a fresh blow. Toward evening Sasha called me to the window.

"What, oh what, has happened to our cigarette?" I sang cheerfully.

But the reply came in his spoken, not sung, voice:

"Genia, you must be brave, there is some bad news. Your husband is here. He was arrested a week ago."

I collapsed onto the bunk. Even today I cannot write calmly of that moment. From the minute I was arrested I had strictly forbidden myself to think about the children. The thought of them drained me of courage. The worst thing of all was to dwell on the details of their daily life.

Vasya had liked falling asleep in my arms and always asked me before he dropped off: "Mother, will you wrap my feet up in the red shawl?" What did the shawl, lying useless on the sofa, mean to him now?

Alyosha and Mayka were always complaining of him and teasing him: "Piggie! Mummy's pet! Sneak!" Sometimes Vasya called me up at work: "Is that the 'versity? Please call Mother."

How well Vera Inber * writes of that "zone of memory" where "the lightest touch wounds mortally." Up to now I had kept such memories at bay by telling myself: "The children are with their father." Foolishly, I had thought our family would be spared. I knew from the prison grapevine that Paul had been dismissed from his post as chairman of the town council, but he had not been expelled from the Party, and he had even been put in charge of the building of a new opera house. This had seemed a good

* Soviet poetess, born 1890, known for her verses on the siege of Leningrad during World War II.

sign. Other people nowadays were swept straight off to prison, without first being downgraded or losing their jobs. But of course it had been absurd to look for rhyme or reason in the acts of lunatics.

Night fell—pitch-black, stifling, smelling of the latrine and the sweat of crowded, unwashed bodies, threaded through with the groans and cries of sleeping men and women, brimful of despair.

It was no use trying to think of things "on a world scale." Tonight I cared nothing for the world, only for my children, orphaned twice over—helpless, small, trusting children, brought up to believe in human kindness. Vasya had once asked me: "Mother, what's the fiercest of all animals?" Fool that I was! Why didn't I tell him the "fiercest" was man—of all animals the one to beware of most.

I no longer struggled against despair, and it clawed its way into me. A special torment was the memory of a trifling incident shortly before my arrest. Vasya had wandered into my room, pulled a bottle of expensive scent off the dressing table, and broken it. I found him picking up the bits and reeking of scent. He looked up sheepishly and said, trying to laugh it off: "I just opened the door, and it fell over."—"Don't tell lies, you nasty child," I snapped at him and gave him a hard slap. He burst into tears.

Now this memory tortured me like fire. Nothing seemed to weigh more heavily on my conscience. My poor little one, alone in this dreadful world—and what did he have to remember his mother by? That she had slapped him for an idiotic bottle of scent. And, worst of all, there was nothing in the world I could do now to put it right.

My pain that night was so great that it brimmed over into the future and reaches me today when I write of it after twenty years. But I must force myself to write. As

Vera Inber says, we must, "without self-pity or indul-
gence, tread these minefields of the mind."

I could never hit the nail on the head as she does. I don't
suppose our nights in prison were worse than hers and her
companions' in the darkness of the siege of Leningrad. But
however senseless and unspeakable their suffering, they
felt that they were fighting fascism; we, tormented in mind
and fettered in body, had not even that consolation. As I
lay, neither asleep nor awake, I felt as though some in-
carnate Evil, almost mystic in its irrationality, were grimacing at me. It seemed as though monsters out of Goya's
imagination were advancing upon me.

I sat up and looked around. Everyone was asleep except
Lydia, who had left her bunk and stood beside me, her
obsessed eyes for once looking at me with simple human
kindness. She stroked my head and repeated several times,
in German, the words of Job: "For the thing which I
greatly feared is upon me, and that which I was afraid of
is come unto me."

This broke the spell. All night I had wanted to cry and
couldn't. My eyes and heart were burning in arid sorrow.
Now I fell sobbing into the arms of this strange woman
from a world unknown to me. She stroked my hair and
said again and again in German: "God protects the father-
less. God is on their side."

22

• *Tukhachevsky and others*

We had noticed for some time that early in the morning, in very clear weather, we could hear snatches of broadcasts from the street through our broken window. The loud-speaker was evidently somewhere nearby, and the wooden screens amplified the sound.

One quiet summer morning we clearly heard the words "Red Army" and "armed forces" being repeated in dramatic tones in connection with the phrase "enemies of the people."

"Something's up again," mumbled Big Anna, rubbing her eyes. "I seem to have missed a lot, not going in for politics. Every day there's some new stunt or other."

"Whenever there's trouble in the army, it means that the foundations of the state are shaken," Derkovskaya announced excitedly.

"Yes, you'd better call a constituent assembly,* " Little Anna snapped at her. Out of sight of the others, she squeezed my hand and whispered anxiously: "Can there really be traitors in the army as well?"

We all stood by the window as if spellbound, but the wind brought only tantalizing fragments—something about "vigilance" and something else that sounded like "treason"; then—as though to spite us, very clearly—"you have just heard," followed by a crackling noise and a military march.

* The last freely elected body in Russia, before it was broken up by the Bolsheviks at its first and only meeting in 1918, had a majority of Social Revolutionaries.

What could have happened? We tapped on the walls, but our neighbors were as bewildered as we were. Not until evening did we get more news. This happened in the following circumstances.

At the hottest hour of the day, when we were lying about in nothing but our pants and brassières, half dead from the heat and stench, the door was opened by the young warder known as Handsome, who mumbled cheerfully:

"Come on, girls, make room for one more!"

We protested. It was not possible. There were already seven of us in a cell for three. Derkovskaya threatened to go on a hunger strike. But Handsome, unversed in the history of revolutionary movement, only added with a grin, "The more the merrier," gave the newcomer a push, and locked the door behind her.

She stood still, framed in the doorway. It took me a few moments to recognize in that mask of horror the face of Zinaida Abramova, the wife of the head of the Council of People's Commissars of the Tartar Republic. He was also a member of the Party's Central Committee and of the presidium of the Central Executive Committee of the USSR. So these were the people they were getting now!

"Zina!"

Certainly no one could have guessed that this petrified woman had yesterday been queening it as "Mrs. Prime Minister." Now she was much more like the Tartar peasant girl, selling cigarettes in a village shop, whom Abramov had married twenty years ago. All the make-believe had been wiped off her face by terror, leaving the signs of her class and nationality plain for all to see. The Tartar accent she had been at such pains to discard sounded clearly in her first few words:

"No, no, I've only been left here for a minute!"

"Just for a minute, is it?" Big Anna said savagely. "Then I needn't get off my bunk. You can stand for a bit."

The sarcasm was lost on her:

"Yes, of course, that's all right. I'll stand."

I had never had any great liking for Zinaida Mikhailovna with her high-and-mighty ways. Her husband, though he often had a drop too much to drink and was running to fat in his office chair, had remained a decent enough fellow who still remembered his working-class past. But Zina, having cast off her former name, Bibi-Zyamal, had broken every link with her village childhood. Dresses, receptions, seaside resorts, filled the whole of her mind and time. Her smiles were rationed in accordance with the social rank of their recipient. It is true that I came in for a little more than was warranted by my husband's status, but this was because Zina was full of respect for the printed word. She liked occasionally to write articles for *Red Tartary* or the weekly *Worker*, and for this she needed my help.

But all this was beside the point now. The woman who stood in the doorway, shocked and almost demented with terror, had to be soothed and comforted as far as lay in our power. I remembered vividly how much Lyama's kindness had helped me in my first days in prison. Going up to Zina, I kissed her and took her in my arms.

"Don't worry, Zina. Come and lie down on my bunk for a bit. Afterward we'll see where to put you."

To my amazement, Zina reacted to my kiss as to a serpent's bite. With a wild shriek, she leaped away from the door, nearly overturning the slop pail. It flashed through my mind that she really was demented, but her next words cleared up the mystery:

"There's a peephole in the door. The warder can see

... He'll think we're old friends. And you ... they wrote about you in the paper. ..."

The whole cell was up in arms at once. "Now you see the moral tone among the members of your Party!" Derkovskaya exclaimed dramatically.

"So you believe what the papers say, do you?" Ira asked, her eyes narrowed. "They wrote that I belonged to the 'right-wing deviation,' but I've never had anything to do with politics, and until I got here I didn't even know what 'deviation' was."

"Well, anyway, Madame is sure to have the best sofa in Vevers's office reserved for her, so we don't have to move up on the bunk." Big Anna lay down and ostentatiously turned her face to the wall.

Zina stood against the door, as though crucified, for about three hours. No one offered to make room for her, and she didn't look as though she would have accepted if anyone had. From time to time she raised herself on tiptoe and looked around in disgust as though afraid to touch anything. In the dingy cell, her snow-white blouse looked like a sea gull unaccountably perched on the edge of a cesspool. At last they came for her. Her face lit up with ecstatic joy—they hadn't let her down. They told her: "We'll need you for a couple of hours." She even smiled at us in farewell.

"Silly bitch!" said Big Anna tersely. "She really thinks they are letting her out! Still, we'll have to think of where we can put her. There isn't room for a sparrow on the bunks, let alone a fat old bag like her."

Several hours passed. Coming back from the washroom in the evening, we heard someone groaning in our cell. On the floor, beside the slop pail, lay Zina. Her white blouse, crumpled, torn, and blood-stained, now looked like a wounded bird. There was a huge bruise on her bare shoulder.

We stared in horror. So it had begun! This was the first case (at any rate, the first we had seen) of a woman being beaten during interrogation.

She was only half conscious and made no reply to our questions. There was not room enough for us to lift her heavy frame onto a bunk. We put a wet towel over her forehead and went silently to bed.

"Genia dear!" I suddenly heard from the corner where Zina lay. Her Tartar accent made it sound like "Ginnia."

"Don't go to sleep, Genia darling, I'm frightened. Tell me, will they shoot us?"

To this day, I cannot forgive myself for the petty spite which made me answer:

"Aren't you afraid to talk to me, after all they wrote about me in the papers?"

I felt wretched the moment I had spoken. Her lips, bruised and split by a man's fist, trembled like a hurt child's.

"Come and lie on my bunk, Zina. I'll sit near you for a while, and you must try to calm down and think things out. What becomes of us will depend on what goes on in the world outside. You were still free this morning. Tell me what the radio said. What's this trouble with the army?"

"Oh, Genia, I'm so frightened! We're not supposed to talk about it here, Genia. All right, then. Don't go. I'll tell you, but only you. Tukhachevsky has turned out to be . . ."

"Yes, and who else?"

But she was again in the grip of panic fear. Instead of answering, she clutched at my fingers and moaned:

"They'll shoot us. I know they will."

Big Anna, meanwhile, had waked up and was sitting up on her bunk. From under the straw pillow she drew Lydia's shiny spectacle case, which served her for a mirror. The first thing she always did on waking up was to look at herself, opening her eyes wide and rubbing the corners, baring her teeth and examining them, pulling at her mop of hair, trying to restore the permanent wave.

"That's fine!" she said, yawning. "We've got an understudy for Nina. They might rehearse a duet. Nina in her contralto: 'When will it all end?' And her majesty, here, squeaking: 'Genia, they'll shoot us.' Then both together: 'Oh, what a pair of dreadful bores we are!' "

I felt genuinely sorry for Zina. I also felt sick at the thought that some ruffian like Tsarevsky or Vevers had smashed his fist into the face of this forty-year-old woman, the mother of two children. But even stronger than pity was my desire to know what had happened that morning in the country, in the army, in our crazy prison life. So I answered Zina with cold calculation:

"If you want me to tell you what is likely to happen to us, I must first know what is going on outside. Tell me, who has been arrested, besides Tukhachevsky, and what for? Then I'll understand the scale of events, and have some idea of whether we, personally, are likely to survive or be killed."

"Genia darling, how can I tell you? The warder is listening. He'll say I'm giving information to prisoners. Then we'll all be worse off."

She got off the bunk and, groaning, lay down on the bare floor, next to the slop pail.

"Go to sleep, Genia!" Big Anna growled. "What on earth do you want to bother with her for? The men will know by tomorrow and they'll sing us the news through the window."

But I had hardly closed my eyes when Zina raised herself painfully and sat up on the floor. She looked ghastly. Bruised and swollen, classless, ageless, almost sexless, she was a groaning, blood-stained piece of flesh.

"I'm frightened, Genia darling. You're educated, you've been to the university, you can tell me, are they going to shoot us or not?"

"Look here, woman," Derkovskaya broke in indignantly, "if you're so concerned for your precious skin why did you go in for politics in the first place? And why do you turn for moral support to people you don't trust? When Genia, who is your fellow Party member, tried to be kind to you, you insulted her, and you still don't even want to tell her what's going on outside. She needs to know. She's been in prison for over four months."

Zina brushed her aside like a buzzing gnat.

"Shut up, Grandma. What are you in for, anyway? I suppose you're religious. One of those pilgrims, I shouldn't wonder."

Derkovskaya smiled scornfully.

"I must have acquired a granddaughter! My name is Derkovskaya. I'm a member of the regional committee of the Social Revolutionary Party."

"Regional committee my foot! I know every blessed member of that committee. And you don't sound like an Old Bolshevik, either. You don't talk right, somehow."

No, we had not found the right way to talk to Zina. Squatting next to where the former First Lady of Tartary was lying beside the slop bucket, I searched my memory and managed to put together a few sentences in Tartar:

"Don't worry, go to sleep. Don't be afraid of me. It's

not true, you know, what they said about me in the papers. Now they'll be writing the same things about you. I'll tell you more tomorrow, and you'll tell me what you can."

I stroked her hair and spoke the names of her children, Remik and Alechka, telling her she must take care of herself for their sake.

Yes, this was the right approach. Zina wiped her hideous, swollen face with the wet towel, and suddenly, in a quick excited whisper, started telling me in Tartar all I wanted to know.

By now she realized that what the papers had said about me was a lie. Hadn't the interrogators cooked up a charge of "bourgeois nationalism" against her and told her that her husband was a Turkish spy? As for the broadcast that morning—no one could understand a thing. Tukhachevsky, Gamarnik, Uborevich, Yakir, and many others—all heads of regional commands—had been arrested. What could it all mean? In Kazan, too, all the senior officers had disappeared, as well as the chairman of the executive committee, the town clerk, and almost the whole of the regional committee.

This was all she knew, but it was plenty to go on with. She stopped talking and looked around. The sight of her torn dress and the cockroaches scurrying about on the floor seemed suddenly to bring home to her the full horror of her position.

"Oh, Genia darling, if you only knew the sort of beds I've slept on in my time!" She closed her eyes, evidently seeing visions of imperial suites at the Moskva Hotel and in government rest homes.

Sleep, poor Zina! You no more deserved those royal suites then, than this filthy cell with its cockroaches and its slop pail now. You should have remained the round-faced peasant girl, Bibi-Zyamal, mowing hay and baking bread. But no, they had first to make you the local Madame Pompadour and then throw you onto this dung heap.

And you and I and all of us will figure in history under the anonymous heading "and others" . . . "Bukharin, Rykov, and others" or "Tukhachevsky, Gamarnik, and others."

Talk about meaning! What wouldn't I have given in those days to understand the meaning of what was going on?

23

• *To Moscow*

The prison hummed like a beehive. It seemed as if the walls were quivering and at any moment might give way under the pressure of the astounding news transmitted through them.

"The whole government of Tartary is in prison."

"The interrogators have been given permission to use physical torture."

"It's the same in Irkutsk, all the leaders are under arrest."

We in Kazan took a special interest in Irkutsk because Razumov, our former regional committee secretary, had been transferred to an equivalent position there and had taken a retinue of Kazan officials with him. He had often

asked my husband and me to join him, and was very hurt
by our refusal. When I ran into him during my troubles
in Moscow, shortly before my arrest, he said with a
triumphant air:

"See what it's like without the boss? If you'd been with
me, do you think I'd have let them treat you like this?"

I had now spent two months in the ancient prison without ever having been called out for questioning. I was the
more alarmed when, the day after Zina's arrival, I was
ordered to get ready for an interrogation at Black Lake.

All the talk around me was of the new system of beating
and torture. Was I destined to go through this as well?
Reading my thoughts, Derkovskaya firmly assured me:

"You've got absolutely nothing to worry about. To
begin with, it's two in the afternoon and the sun's shining—
they keep all such things for nighttime. And secondly,
your case is closed. They're most likely summoning you
just to tell you that."

So it turned out. I had been summoned to sign a statement to the effect that the investigation was closed, that
my offenses came under Article 58, sections 8 and 11, of
the legal code, and that my case was being referred to the
military tribunal of the Supreme Court. This was told me
by Bikchentayev.

He was in excellent spirits. The sun shone on his water
jug and his polished boot-button eyes. Writing assiduously
and handing the document over to me to sign, he looked
up now and again with cheerful expectancy, as if seeking
my approval for his diligence. I felt I should admire the
deftness and efficiency with which he was conducting my
affairs.

"So," he finally announced in a good-natured tone,
"your case will be heard by the military tribunal of the

Supreme Court of the USSR. That means you soon get sent to Moscow."

He looked up again, as if wondering why I was not pleased by the news. Determined to get some sort of reaction out of me, he added: "You've nothing to complain of. I've been very fair. You knew Japanese spy Razumov, I could make big charge of this but I left it out."

"Who's that? What Japanese spy are you talking about? Is that the regional committee secretary of Irkutsk? The one who's been a Party member since 1912 and belongs to the Central Committee?"

"Yes, spy Razumov evaded Party vigilance and rose to important post. He did join Party before the revolution—he was agent of Tsarist secret police."

How often had Tsarevsky and Vevers threatened to charge me with "attempting to discredit the regional committee in the person of its former secretary Razumov"? In vain I had assured them that Razumov and I were friends, and that what they called my efforts to discredit him had been nothing more than friendly arguments between us. They stuck to their point, though they did not bring a formal charge: they had more than enough "evidence" to send me before the military tribunal without that, and now . . .

Now it seemed that, far from being in need of protection from terrorist plots, regional committee secretaries were themselves the leaders of such conspiracies! We already knew that a schoolboy of sixteen was held in our prison on the charge of plotting the murder of regional committee secretary Lepa. Now Lepa and all the other heads of committees were in jail.

My news about being sent to Moscow caused a sensation in the cell. My companions looked at one another in silence. Finally Derkovskaya said:

"Did he tell you what section 8 means?"

"No. I didn't ask. Does it matter?"

"Yes, it does. It means that you are being charged with terrorism. And section 11 applies to groups. Membership in a terrorist group. It's no joke. And you are being sent before a military tribunal."

I have often thought since then that my behavior at the time must have struck my cellmates as highly courageous. In fact, it was not courage but lack of imagination. I could not believe that I was really in danger of being sentenced to death. Quite unaccountably, I failed either to hear, or to take in, Derkovskaya's explanation that the minimum penalty imposed under sections 8 and 11 was ten years of hard labor. If this was the minimum, I could not help knowing that the maximum was death.

The full terror of death did not come over me till much later, when I was in the Lefortovo prison in Moscow. Here in Kazan, amidst the flood of fantastic news which seemed to suggest that there had been a *coup d'état*, everything appeared to be part of the same senseless, unbelievable turmoil. At any moment, it seemed, the Party—that section of it which was still at liberty—would seize the madman's hand and say: "Enough! Let's first decide who among us is the traitor!"

All my cellmates, whatever their political views, gave me a warm and loving send-off. They sewed on buttons for me, darned my stockings, and plied me with advice and the addresses of their relatives. Tortured by a single thought, I heard it all as in a dream.

Derkovskaya had told me that prisoners about to be deported were allowed to see their family. I pictured to myself my mother's eyes and the frightened, puzzled faces of the children as they looked at me through a grating. Must this happen? The memory might torment them for the rest of their lives.

But I need not have worried. Derkovskaya had spoken out of her experience of Tsarist prisons. There was no room nowadays for "rotten liberalism" or "pseudo-humanitarianism." I was not allowed to see my family, and I never saw Alyosha again, or my mother.

24

• *Transfer*

"Bring your things!"

What a lot of meaning was contained in these few words! Once again you are bound to the wheel, spinning and dragging you away at your master's will from everything that's near and dear, toward the unknown.

Lydia, the Seventh-Day Adventist, for once took advantage of the opportunity to preach her faith:

"We are all grains of sand blown by the wind. This trial has been sent you so that you should realize in whose hands your future lies."

"But if the hands are those of scoundrels like Tsarevsky, it's too shameful. It's wrong to submit to them. I ought to put an end to it all, but I haven't yet the strength."

"God preserve you from doing such a thing! You'd be killing your soul."

Forgetting her principles as a Social Revolutionary, and her irreconcilable hostility to all Communists, Derkovskaya wiped away her tears.

"We'll miss you in the cell. There won't be anyone to recite poetry. I'd never realized before that Blok * was such a splendid poet."

"I'm surprised at you," I said jokingly, "crying over me without asking Mukhina's permission. Would she allow you to weep over a Communist who's not in the opposition?"

Derkovskaya shrugged her shoulders crossly and blew her nose on a towel. By way of farewell, I recited Mandelshtam's † melancholy poem:

The horses tread slowly,
The lamps burn low,
And where they are taking me
Only strangers know.

A fairly large number of people were being sent to Moscow for the session of the military tribunal. Our sharpened hearing told us that they were being collected from several cells, and from ours Ira was taken as well as myself. This struck her and many others as unfair.

"It's all very well their taking you," she said, "but why on earth should I be dragged before a military tribunal?"

The idea that membership in the Communist Party was bound to aggravate the defendant's case had by now taken

---

* Alexander Blok, 1880–1921, famous Symbolist poet, best known as the author of *The Twelve*.
† Osip Mandelshtam, 1892–1938, famous Russian poet who disappeared in the purges.

firm root in everyone's mind. How indeed could the situation be explained except as Stalin's death blow to the Party, his "eighteenth Brumaire"?

At last we were ready, our bundles tied, our farewells said, our last messages exchanged by wall telegraph or by operatic radio. Ira and I sat waiting on the plank bed, but our thoughts were already very far away. As in a dream I heard Zina complaining:

"You're all right, Genia dear! You've got your higher education, you'll always fall on your feet. But what about me? . . ."

If only she had known how little use I was to find for higher education and how much for physical endurance!

The door opened. We were taken out into the corridor and down the stairs. But what was this? Another mistake of the guards, as on the day we arrived? Sitting on their bundles at the wooden gate with its iron cross-bars were two women whom I knew well from my university days— Julia Karepova, a biologist, and Rimma Faridova, a historian.

No, it was no mistake. We had been brought together deliberately. We were all being taken to Moscow. We questioned one another eagerly. It appeared that Julia and Ira, as former members of Slepkov's seminar, were involved in the same case. Now that the investigation was over, they could safely be allowed to meet. My assumption that as a former student of Elvov's, Rimma had been similarly paired with me, turned out to be wrong.

"No," she cheerfully explained. "I'm a Tartar, so it was simpler to put me down as a bourgeois nationalist. Actually, they did classify me as a Trotskyist at first, but Rud sent the file back, saying they'd exceeded the quota for Trotskyists but were behind on nationalists, even though they'd pulled in all the Tartar writers they could think of."

She used all these exotic terms without the slightest irony, just as though she were discussing the fulfillment of an ordinary economic plan. Nothing about it seemed to strike her as odd, and unlike the rest of us she looked the picture of health. I was later to discover the reason: from the very start, she had given the interrogators everything they wanted. Dozens of Tartar intellectuals and Party activists were sacrificed on the altar of her relative well-being in prison. One of them was her husband, formerly the head of the department of culture and propaganda in the regional committee. A reserved, intelligent, silent man who looked like a Chinese, his nickname at "Livadia" had been Confucius. It was the evidence supplied by his wife which had sent him to his death. She was rewarded with thirty genuine pieces of silver in the form of permission to get any number of parcels, and thirty false ones—a promise that she would not be sent to prison or to camp, but only into exile, and only for a maximum of three years at that.

Julia amazed me by her description of the behavior of Slepkov. Brought back for "re-investigation" from Ufa, where he had spent three years in a security prison for political criminals, he too had "fully co-operated" and supplied the names of over a hundred and fifty of his "recruits." Always willing to "confront" the defendants, he had taken part in a confrontation with Julia. This she described as a gruesome affair in which both Slepkov and the interrogator had behaved like amateur actors who speak their lines without the least attempt to carry conviction.

Looking her blankly in the face, Slepkov stated that,

having received "terrorist instructions from Bukharin in Moscow," he had come to Kazan to communicate them to the leaders of an underground center, among them Julia, who had at once expressed her willingness to undertake a number of terrorist acts. Half choking with amazement and anger, she shouted "Liar!," to which Slepkov replied with pathos: "We must disarm! We must go down on our knees to the Party!"

As a result, Julia's case looked a great deal better documented than mine. Instead of "witnesses" who professed to know of the existence of a subversive group but not to have belonged to it themselves, the authorities held the alleged "leader of the Bukharinist underground in Kazan," who in his "confession" had "unmasked a member of his group"—poor little round-eyed Julia, the most orthodox of all the Party members I had ever known.

To this day, I cannot understand what could have made Slepkov act as he did. He had always seemed a captivating person whose attraction lay not only in his brilliance as a scholar but in his kindliness as a man.

Could it really have been a sordid attempt to buy his own life at the cost of hundreds of others? Or was it perhaps the very same tactic which Garey had once urged upon me: to sign everything in the hope of reducing the situation to such absurdity as to provoke an outburst of indignation within the Party? Whatever his reasons, they were no more unaccountable than many other things in the fantastic world in which I was condemned to live, and soon perhaps to die.

All four of us were to appear before a military tribunal on a charge of political terrorism. Rimma assured us that we were the only women sent on this grim journey, along with a multitude of men.

All this ought surely to have made us face the likelihood of our being sentenced to death. It would have been only logical. But evidently we, too, must have been influenced by the law of the reversal of logic which ruled in our crazy world, for not one of us would admit the possibility for a moment.

Ira kept harping on the fact that she was not a Party member, which, in her opinion, gave her an enormous advantage over the three of us Communists. Rimma pinned her faith on the interrogator's promise of "exile, not prison," while Julia and I found comfort in the gigantic scale of the current operation—"you can't shoot everyone"—and for some reason in the fate of Zinovyev, Kamenev, and Radek: "If they got only ten years each, how could we possibly get more?" * The simplicity of our reasoning can perhaps be excused on the grounds that we had been in prison for six months and had not therefore had the chance to watch, day after day, the gruesome progress of what, since Stalin's death, is referred to by the academic term "violation of socialist legality."

The gates of the ancient prison clanged behind us. The Black Maria had already been filled up. Coughs and sighs came from inside the cells. As the four of us need no longer be concealed from one another, we sat on our bundles in the narrow passage. We could see a little through a crack in the rear door—not, of course, with our ordinary vision but with our practiced, intensely observant and sensitized prisoners' eyes. The senses of smell, hearing, and direction helped as well.

The scent of lime trees drifted in. It meant that we were passing the Lobachevsky monument. A big pothole in the

---

* Actually, they had already been executed.

asphalt of the roadway—we were turning into Small Pro-
lomnaya Street. Imagination supplied the rest, engraving
on my mind the memory of the beloved city, my second
home, where "I had suffered, loved, and buried my heart."
Sentimental, no doubt, but at such a moment one is entitled to be.

We stopped. There was a smell of hot rails and engine
soot, the sound of engines busily puffing, then of their
brief, nervous whistles.

"Everybody out!"

It was not the familiar station. We were somewhere at
a halfway halt. And what about my meeting with my
mother, my children? Hadn't Bikchentayev promised? No,
there was no one on the platform but a large covey of
interrogators and guards. After the darkness in the Black
Maria, our eyes were dazzled by insignia, brass buttons,
a few medals. Against this background, Vevers stood out
in his elegant, dove-gray civilian suit. His face was attentive, strained, yet wore the familiar grimace of hatred and
contempt which they were taught to practice at their
training schools.

"This way! This way!"

It was an ordinary third-class coach divided into com-
partments, each seating four. There was a guard at the door
of each compartment, except the one in the middle, reserved for the interrogators in charge of the valuable
human freight they were escorting to Moscow.

A jolt as the engine was coupled on. Another jolt and
we were off. . . . I knew what I was leaving behind—my
children abandoned to the mercy of fate (if only it were
fate! The NKVD was much more frightening!), my
mother, my university, my books, a clean and happy life
full of the consciousness of having chosen the right way.

As for our destination—only those who were taking us there could tell. . . .

Tsarevsky came in and gave us our instructions for the journey: how to eat, drink, sleep, and go to the lavatory. It was a long time since I had seen him and I noticed something new about his face. It was earthy-gray by contrast with his straw-colored hair and looked old, although he could not have been more than thirty-five. His voice was still the same—grating, vulgar, jeering—but in his eyes there was now something more than meanness, namely fear.

We could not understand this at the time. Later we were to learn that the purge of the NKVD itself had already begun. "The Moor has done his duty—the Moor can go." A number of interrogators had already been singled out, and, as experts in such matters, they were dimly aware of this. Tsarevsky was to be arrested shortly after our dispatch to Moscow; soon after, he hanged himself in his cell with a strap he had managed to conceal. It was said that he had signaled to his neighbors through the wall, advising them to "sign nothing."

Bikchentayev, the cheerful turkey cock, was to get fifteen years. Rud was among the victims, and so was Yelshin, whom I later met at Kolyma. But all this was still ahead. Meanwhile, after telling us in detail all we were forbidden to do, Tsarevsky went to his compartment to a dinner of pork chops—the smell drifted down the passage—and white wine; the guards carried out the bottles.

Our window was thickly smeared with white paint. Only when we went to the lavatory did we occasionally catch sight, through the half-open door of the platform at the end of the carriage, of some well-remembered landmark on the familiar Kazan-Moscow route.

One incident has stuck in my mind. We noticed at a station where we stopped that the guards were passing small baskets of fresh raspberries from hand to hand. Ira had with her fifty rubles which her father, a professor at the Institute of Civil Engineering, had succeeded, by hook or by crook, in sending her.

"Would you ask Interrogator Tsarevsky to come in for a moment?" she asked the guard and, when Tsarevsky appeared, turned to him with the air of a coquettish woman of the world:

"Would you be so kind, Lieutenant, as to tell your men to buy us some raspberries? Here's the money. . . ."

Whether Tsarevsky was moved by the absurd contrast between her tone and our surroundings, or by obscure forebodings of his own fate, he took the money and in a few minutes came back with two small boxes of raspberries and the change. The raspberries were very fresh, covered with a silvery bloom, and so fragrant that it seemed a shame to eat them. We counted them, divided them into four equal shares, and spent an hour and a half eating them one by one, laughing with happiness.

We had managed to snatch one last morsel from the banquet of life from which we were about to be excluded for ever.

25

• *Introduction to Butyrki*

From the moment we arrived in Moscow we could sense the tremendous scale of the operation in which we were involved as victims. The various agencies concerned with carrying it out were inhumanly overworked. People were run off their feet, transport was insufficient, the cells were full to bursting, the courts sat for twenty-four hours a day.

Long after our train reached the station we were kept in the carriage listening to the sound of feet running along the platform, abrupt shouts, mysterious creaking and clanking. . . . At last we were loaded into a Black Maria. From the outside it looked bigger and handsomer than the one in Kazan. It was painted light blue, and passers-by no doubt imagined that it was carrying loaves of bread, milk, or sausages. But the cages inside were even more unendurably stuffy than those at Kazan. They were coated with oil paint and completely airless, so that within minutes on that scorching July day with its smell of melting asphalt, we felt suffocated.

Exhausted and dripping with sweat, our hair matted, we sat gasping in our cages, patiently waiting. It was a long wait—apparently they were short of drivers. Meanwhile the same sounds of running, whispering, knocking, doors slamming. . . . Certainly they were not having an easy time out there.

Guessing by some sixth sense that there was as yet no guard in our van, we began to talk. It turned out that all the occupants were from Kazan but that the only women were the four of us. Among the men were almost the whole of the former government of the Tartar Republic and several members of the regional committee board. Abdullin, who was soon to be executed, was there too, and we managed to exchange a last farewell with him.

All of a sudden the tramp of boots came closer; the doors banged, the engine roared, and we were off. The drive was a long one. That meant that we were going to the Butyrki prison: the Lubyanka was only a short way from the Kazan station. The air in the van got worse than ever. Someone shouted: "Open the door, I'm going to be sick." "Against orders" was the curt reply. My hands and feet went numb, my brain clouded, strange pictures passed in front of me. I remembered reading that the victims of the French revolution were taken to the guillotine in open tumbrils, and not stifled to death on the way: according to Anatole France, César Birotteau had even read Lucretius, standing up, till the very last moment. In order not to lose consciousness I tried to occupy my mind by picturing the streets we were passing through. Then everything went black before my eyes.

A strong smell of liquid ammonia brought me to myself. The van had stopped, the door of my murder chamber was opened, and someone in a white smock pushed a phial under my nose; then they opened the other doors, one by one, and repeated the process. Evidently the drive had been too much for the male prisoners also.

I must have walked from the van into Butyrki in a semiconscious condition, for I can remember nothing about it. When I came to myself I was sitting on my bundle of clothes in an enormous hall which was not unlike a railway station. Vast and echoing, it was fairly clean, with men and women bustling about in uniforms that were not unlike those of railway employees. There were a lot of doors, and some windowless cabins resembling telephone booths. I learned later that they were the so-called "kennels" into which prisoners were put when they had to wait for any reason. The basic law of prison was strict isolation: none of us was allowed to see anyone except his cellmates. A bell at the door announced the arrival of a new group of prisoners. A wardress came up and said to me quietly: "Over there." In a moment I was in a kennel, locked in and once more by myself. The place was a little larger than a telephone booth, tiled on the inside, with a stool to sit on and a ceiling light. I had hardly had time to look around when the lock clicked again and I was taken out into a large room full of naked or half-dressed women. The wardresses in their black jackets looked like so many jackdaws.

Was this a bath or a medical inspection? No, it was for the purpose of searching the new arrivals.

"Take your clothes off. Let your hair down. Spread your fingers and toes. Open your mouths. Stand with legs apart."

The wardresses, with stony faces and quick, efficient movements, ran their fingers through our hair as if hunting lice, examined our mouths and anuses. Some of the prisoners looked terrified, others disgusted. A great many of them obviously belonged to the educated classes.

The work proceeded swiftly. A pile of confiscated objects soon covered the long table: brooches, rings, watches, earrings, garters, notebooks. These were clearly Moscow women, arrested only today and brought straight from home with all sorts of knickknacks on their persons. They were worse off than I was: I had the advantage of six months' experience behind me, plus the fact that I had nothing left to lose.

"Put your clothes on again!"

A young girl came up to me—almost a child in appearance, with her hair cropped in boyish fashion—and said, "Excuse me, but are you a Party member? You look like one—it seems a funny thing to ask here, but I must know. You are? Well, I belong to the Komsomol, my name's Katya Shirokova, and I'm eighteen years old. I need some advice. You see, that German woman over there has hidden some gold things in her hair and I can't decide whether I ought to tell the wardress. I don't like to give her away, it's disgusting, but on the other hand this is a Soviet prison and for all I know she may be a real class enemy."

"What about you and me, Katya? Are we?"

"Oh, of course it's a mistake about us. You can't make an omelette without breaking eggs. Anyway, they're sure to let us out. But meanwhile it's terribly hard to know what to do, especially about her."

I looked at the woman at whom Katya was pointing, and saw a face of unusually tender beauty and charm. I found out later that she was a well-known German film actress, Carola Heintschke. She had come to the Soviet Union in 1934 with her husband, who was an engineer. The two earrings which she had skillfully concealed from the watchful eyes of the wardresses were a present from her husband, whom she believed to be already dead. With the deft movement of an actress used to playing in adventure films, she had managed to hide them in her luxuriant blonde hair.

Katya's quaint, attractive face was still turned to me in earnest inquiry.

"So you want a directive, Katya, do you?"

"Well, it is difficult, you know, she's a German and . . ."

"Listen to me. Since we're all naked in every sense of the word, I think you should be guided by the instinct which is generally known as conscience. And I imagine yours tells you that it would be a dirty trick to give the woman away."

Thus Carola Heintschke's earrings were saved—like herself, for a short time only. But more of that later.

The formalities went on till late into the night. After the search they took our fingerprints, an equally humiliating procedure. Then we were photographed full-face and in profile, and at last came the long-awaited bath, a joy in itself and an interlude of sanity in this Dantesque Inferno.

There is no place where people become acquainted more quickly than in prison, especially during such "processing," under the influence of a common fear of the immediate future and a shared sense of outraged human dignity. We forty women went through all the stages together, from the morning search onward, awaiting our turn together, telling each other in brief, frightened whispers what we had been arrested for, the names of our children, our griefs and injuries. I already felt that life would be much more bearable if I were not parted from black-haired Zoya, of Moscow Medical School, about whom I had learned in those few hours as much as one usually learns in ten years of close friendship. She evidently felt the same, and when I came out of the photographing booth she ran up to me with a sigh of relief and said:

"We're still together, Genia. I'm sure they'll put us in the same cell. It would be marvelous!"

But even these small consolations were denied us: we were separated as if at a slave market. Coming out of the shower room, I saw that neither Zoya, nor Katya Shirokova, nor the golden-haired Carola was in the corridor.

"Left!" the warder commanded, and I was led alone through the dark corridors. Then I was handed over to another warder and heard the whisper "To the special block." Here I was taken in charge by a wardress in a dark jacket and with a severe, monastic face.

The doors in my new abode were ordinary ones, locked by a single key, without medieval bolts and padlocks. I stood with my bundle in my hands, gazing around me. The huge cell was crowded with women; the regular noise of breathing was constantly interrupted by shrieks, groans, and muttering. It was clear that the inmates were not only asleep but were having fearful nightmares. Compared to the two prisons I knew, this one was almost luxurious. It had a large window, with a grating of course, but the screen behind it was not of wood but of frosted glass. There were collapsible beds instead of a single plank bed; the enormous slop bucket in the corner had a tightly fitting lid.

All the beds were occupied. After waiting a little I undid my bundle and took out my cotton blanket (a check one, dear to me because Alyosha had used it) and spread it on the floor near the window. I stretched out my legs rapturously. My body was racked with fatigue. I was about to sink into blissful oblivion when the peephole opened and a wardress stuck her head through.

"You mustn't lie on the floor. Get up!"

"But there's no room anywhere else."

"You can sit till morning—it isn't long. Then you'll be taken to another cell."

As soon as her head disappeared, a tousled figure from one of the bunks addressed me in a Caucasian accent:

"Here, comrade, take my place. I can't sleep anyway. Honestly, I'm quite happy to sit up for a while."

She quickly installed me on her bed. What bliss! I had forgotten what it was like to lie on anything except straw. The pillow smelled of something I had forgotten too—cleanliness, a trace of scent.

My new friend understood my thoughts.

"Yes, in Armenia we had a bit of rotten liberalism; I was allowed to bring a pillow and some linen from home. They wanted to take the pillow away here, but the interrogator stuck up for me. He's trying to get around me at the moment, and thinks I'll sign something."

Whether because I was so tired or for some other reason, the voice seemed familiar. I could not see her face: the light had been switched off, and dawn was only just stealing through the opaque glass and grating.

"Are you all right now? Well, that's marvelous!"

This expression too struck a chord. I shook off my sleep and tried hard to remember: that phrase, that tousled head. I took her by the hand and said, "What's your first name?" She replied "Nushik," and I jumped up and threw my arms around her.

"Nushik! Look at me! Don't you recognize me?"

"Genia! How stupid of me! Of course!"

Laughing and crying and interrupting each other, we talked about the past. Eight years before, when we were young postgraduate students, we had slept in the dormitory at a teachers' training college in Leningrad.

"It was nearly as big as this, wasn't it?"

"Yes, but a little different."

It was in fact a big drawing room in the former palace

of the Grand Duke Sergey Alexandrovich, close to the Hermitage Museum. One wall was entirely taken up by a plate-glass window which looked out on the embankment. Ten of us slept in the room, which was lit up at night by a ghostly glimmer from the riverside lamps.

"Do you remember the time I woke you up in the middle of the night?"

Of course I did! She had spent the day frantically studying for an exam in dialectical materialism, and she had wakened me to ask: "Tell me, darling—who was it that Marx stood on his head? Hegel? M-arvelous!"

We talked on and on, exchanging pleasant memories, and after a while she said:

"I'm going to give you a bit of information in return. Do you know who's stood everything on its head now?"

I had a pretty good idea, but I let her tell me. She whispered into my ear: "Stalin!"

We went on whispering for a long time, and I could not tell at what moment I fell asleep. When I awoke, it was to feel someone's eyes fixed upon me. A woman aged about fifty-five was sitting beside Nushik at the foot of the bed, an expression of acute suffering on her face. Seeing that I was awake, she moved closer to me and asked, wringing her hands:

"Tell me, have they been tried yet? They've been shot, haven't they?"

"Who? What trial?"

"Are you afraid to talk about them?"

"You needn't be afraid, Genia," Nushik put in. "This is Rykov's wife. She wants to know what's happened to her husband. We've been here for two months, and we don't know anything."

I explained as convincingly as I could that I had been in

prison for six months, that I had been brought from another town and knew nothing about Rykov's impending
trial. But his wife refused to believe me, partly because I
had only just arrived and looked fairly fresh after my bath,
but chiefly because people behind bars were no less afraid
of betraying themselves than those outside. Although they
were already in the spider's web, they felt that they might
yet struggle out of it, that their neighbor's offenses might
be worse than their own, that they must be careful not to
give anything away. I was to meet many such "diplomatic"
prisoners, who swore they had not looked at a newspaper
for a year before their arrest, and knew nothing whatsoever. There were many, too, who carried on ultra-patriotic
conversations at the top of their voice, in the naïve hope
that their words would be overheard and reported where
they would do most good.

It was humiliating to be mistaken for one of this sort.
But there was no time to argue. Again the wardress stuck
her head in.

"Get up. Get ready to go to the washroom."

Thirty-eight folding beds creaked. Everybody got up.
I looked eagerly into their faces. Who were they? Those
four, for instance, in absurd, low-cut evening dresses and
high-heeled shoes? All looking bedraggled, of course . . .
What could they be?

Nushik came to my assistance:

"They're not tarts, silly. They're Party members, friends
of Rudzutak.* They were having supper with him after
the theater and were arrested in what they stood up in.
That was three months ago, and the poor things haven't
been allowed any parcels, so they're still in their evening

---

* Disgraced Bolshevik leader.

dresses. I gave the old one a scarf of mine yesterday—clothing the naked, you might say."

All the thirty-eight women dressed as fast as they could, afraid of being late for the washroom. The cell buzzed with low-voiced conversation. Many people were telling their dreams.

"They've nearly all become superstitious," said Nushik. "You see that old girl by the window—she tells her dreams every morning and asks us what they mean, and do you know what she is? She's a professor! . . . Look at that child over there. That's Nina Lugovskaya. She's sixteen. Her father's a Social Revolutionary, he's been in since 1935, and now they've taken his wife and his three girls as well. Nina's the youngest, she's still at school."

Presently, all of us—thirty-nine including myself, from sixteen-year-old Nina to the seventy-four-year-old veteran Bolshevik Surikova—were milling around in the large and not too dirty washroom which looked like a station cloakroom, and we all hurried as if our train were about to leave.

There was a lot to do, including washing our underclothes. This was strictly forbidden, but people took the risk because most of them got no parcels and had only one set of things.

Everyone made much of Nina—washing her panties, combing her hair, giving her extra lumps of sugar and advice on how to behave during her interrogation.

I felt an almost physical pang of acute pity for the young and the old—Katya Shirokova or little Nina, who was hardly older than our Mayka, and Surikova, nearly twenty years older than my mother.

Yes, I was very lucky—lucky to be over thirty, but not much over it. I had my own teeth and I could see without spectacles. (Everyone who wore glasses had had them taken away, and the near- and far-sighted suffered terribly.) My heart and stomach and all my other organs were in perfect order. At the same time I was old enough to be tough, I wouldn't break, I wasn't a reed like the Ninas and the Katyas.

So I had no business to be sorry for myself. I was luckier than most. There was only one thing. It seemed to me that I suffered more than any of us from the humiliation of all they were doing to us. The worst physical sufferings, it seemed to me, would have been easier to bear than the sense of outrage and degradation.

The only way to overcome it was to tell myself at every hour of the day that those who did these things were not human beings. After all, I would not have felt insulted if a monkey or a pig had scrabbled in my hair, looking for "substantive evidence of my crimes."

26

• *The whole of the Comintern*

The wardress stopped me from going back to the cell with the others. She told me to wait and, having locked the door behind them (I did not even have a chance to say good-by to Nushik), led me along the corridor and pointed to an open door.

"In there!"

It was a cell exactly like the one in which I had spent the night, but empty, as the inmates were in the washroom. The wardress indicated my bed: it was near the door, and therefore near the slop bucket. But in other respects I quite liked the look of things. The sun shone dimly through the frosted glass; there were thirty-five collapsible beds, all tidily made, but the main thing was—did my eyes deceive me? No, there were actually books on each one. I trembled with delight. My beloved, inseparable companions whom I had not seen for six months past—six months without leafing through your pages, without smelling the acrid printer's ink! I took up the nearest one—it was Kellermann's *Tunnel* in German. The second was a volume of Stefan Zweig, also in German; then there was Anatole France in French and Dickens in English. Before long, I discovered that all the books were foreign. The ragged and crumpled articles of clothing scattered about on the beds also had a foreign look.

The key turned, the door opened again, and in trooped about thirty-five women, chattering in subdued tones in a mixture of foreign languages. They surrounded me and plied me with questions in a friendly manner, talking in German, French, and scraps of Russian. Who was I, when had I been arrested, and what was happening outside? I replied in Russian and, in my turn, asked:

"What about you, comrades? I can see you're foreigners, but where are you from?"

A slim blonde woman of about twenty-nine, who was standing in front of the group, held out her hand and said in broken Russian:

"I introduce myself—Greta Kästner, member of the German Communist Party. This is my—how you say—

friend Klara. She ran away from Hitler—the Gestapo held her for long time."

Klara was very dark, more like an Italian than a German. She looked at me expectantly and nodded to confirm what Greta had said. Then came another tall blonde, who announced in excellent Russian that she was a member of the Latvian Communist Party. Then a *Communista Italiana*, then a Chinese Party member of uncertain age who put her arms around me and said: "In Russian they call me Genia too—Genia Koverkova. I studied at the Sun Yat-sen University in Moscow, and they gave us all Russian names. But tell us about yourself, comrade."

When they heard that I was a member of the Communist Party of the USSR, they became greatly excited and asked question after question. What were the details of the Red Army plot? Was Wilhelm Pieck * in prison? Had all the Latvian rifle corps † been arrested? When was the trial of Bukharin and Rykov ‡ to begin? Was it true that there had been a July plenum of the Central Committee at which Stalin had demanded that prison regulations be tightened?

I explained that all these matters were new to me, that I had been in prison longer than they and had been brought from a provincial town to be tried by a military tribunal. After a while the cluster of women began to disperse and I was left with the two Germans, Greta and Klara. My German was as full of mistakes as their Russian, but we talked animatedly to our mutual satisfaction in both languages at once.

* German Communist leader, 1876–1960, first head of the East German State.
† Crack unit known for their fanatical devotion to the Soviet regime.
‡ Nikolay Bukharin and Aleksey Rykov, Bolshevik leaders executed in 1938.

"What are you accused of, Greta?"

The "Aryan" blue eyes glistened with unshed tears as she replied:

"Oh, *schrecklich*—espionage."

Then she told me a little about her husband—a true working man. She herself had belonged to the Communist youth movement since she was fifteen. But what had happened to her was nothing, it was Klärchen . . .

Klara lay down on the bed, turned over on her stomach and pulled up her skirt. Her calves and buttocks were covered with deep, hideous scars, as though wild beasts had been clawing at her flesh. Her lips tight in her swarthy face and her gray eyes flashing pale fire, she said hoarsely:

"This is—Gestapo." Then she quickly sat up again and, stretching out both her hands, added: "This is NKVD."

The nails of both her hands were deformed, the fingers blue and swollen. My heart almost stopped beating. What could this mean?

"They have special apparatus to produce—*wie sagt man?* —sincere confession."

"Torture!" I exclaimed.

Greta nodded sadly. "Night come, you hear."

At that moment I heard a Russian voice say: "May I speak to you for a moment, comrade?"

Apparently there were a few Soviet citizens in the cell besides myself. This one was a woman in her late thirties called Julia Annenkova, who had edited a German newspaper in Moscow. She was not beautiful, but her face was striking and she looked as though she was of French descent. Her eyes smoldered mournfully. She took my arm, led me to one side, and whispered confidentially:

"You were quite right not to answer their questions. Who knows which of them really is an enemy, and which are the victims of a mistake, like you and me? I advise you to go on being careful, to make sure that you don't commit a crime against the Party after all. The best way is to say nothing."

"But it's the truth—I don't know anything. I come from the provinces and I've been in prison for six months. Do *you* know what's going on in the country?"

"Treason—appalling treason which has worked its way into every branch of the government and Party organization. Secretaries of territorial committees, secretaries of national minority Communist Parties are among the traitors —Postyshev, Khatayevich, Eiche, Razumov, Ivanov—Antipov, the president of the Soviet Board of Control—lots of army officers . . ."

"But if all these people have betrayed one man, isn't it easier to suppose that he has betrayed them?"

Julia turned pale and, after a moment's silence, said curtly: "I beg your pardon, I was mistaken in you."

She moved away, and I was buttonholed by another Russian, Natasha Stolyarova. She was twenty-two and looked like a schoolgirl with her auburn pigtails and round, freckled face. She had emigrated with her parents at the age of five or six and spent her childhood in Paris, where she became bilingual. A few years earlier she had come back to Moscow as a repatriate. Intoxicated with the Russian atmosphere and the pure Russian speech, she had taken up the career of interpreter. And now she too began to warn me to watch my tongue.

"You're too trusting. Why did you offend Julia like that? You can see how fanatical she is—a person like that can easily be a stool pigeon, and why should you play into the interrogators' hands?"

Natasha assured me that she, with her "fresh eye," could see more than the rest of us.

"Believe me, the Caucasian usurper is even worse than his French predecessors. 'Off with their heads'—that's his answer to everything."

"But can he really be deliberately out to destroy the best elements of the Party? What could he base himself on after that?"

"Just you wait till night comes—you'll hear what he bases himself on."

"But I spent a night in the other cell and I didn't hear anything."

"That's because you didn't arrive till just before dawn. The torturing goes on till three. And the German women, who've been through the hands of the Gestapo, say that the people here must have learned it from them—the style's the same. A spell of foreign training, would you say?"

Natasha's harsh, agonizing words contrasted strangely with her schoolgirl's braids. Specks of sunshine, dimmed by the frosted glass, danced on her reddish hair. In the same way life, light, and human kindness kept breaking through the gloom which surrounded us.

Greta was describing to her friend the ravishing dress which she had worn to the last First of May party at the Bolshoy Theater, and Klara's eyes were sparkling with curiosity. She too was talking about a special dress, drawing its shape in the air, tracing it with her blue fingers with their crushed nails.

As for the Chinese Genia, she was teaching a slim Polish woman named Wanda exercises that were "good for the figure." Looking furtively at the peephole from time to time, the two of them were lying on their backs on the floor and pedaling with their legs in the air, anxious to preserve their figures from the effects of lying about all day, of prison inactivity, of the diet of *kasha* and oatmeal gruel. . . .

Dinner and supper came and went. The evening wash. Inspection. Lights out. We lay in our beds and waited. Soon *it* would begin, inexorable as death.

27

• *Butyrki nights*

That evening, the general mood was more hectic than usual because of an incident during inspection.

The rule at Butyrki was to count us not by heads but by checking our tin mugs. We were supposed to place these on the table before inspection. The warders and block supervisors would count the mugs and leave us with the usual parting instructions about not talking too loud, going to sleep immediately after "lights out," and so forth.

On this occasion the warder who did the counting was exceptionally, unbelievably stupid. He miscounted several times, arranged the mugs in a tidier pattern, lost count again, and had to start from the beginning, licking his right thumb in a comical manner.

The first to giggle was the Chinese, Genia Koverkova, and this set the rest of us off. When the ceremony was over and the senior warders had gravely withdrawn with their attendants, the cell resounded with peals of crazy laughter such as one sometimes hears in prisons. As if to compensate for their anxiety, grief, and sufferings, prisoners will laugh at the slightest provocation. They explode with a Homeric mirth completely disproportionate to what caused it. Such outbursts are not easy to stop, and on this occasion the warnings of the more cautious among us had no effect.

"Be quiet, all of you!" This shrill command came from Julia Annenkova—her face pale and contorted, her hand lifted in a threatening gesture. "How dare you make fun of him! He is doing his duty as the representative of Soviet power. How dare you, how dare you!"

We stopped laughing abruptly. A tall, earnest, German woman named Erna tried to explain to Julia that our amusement was provoked "by the comic character of the man himself, irrespective of his social functions." We should have laughed at him in the same way if he had not been a warder, but a prisoner like ourselves.

A voice from a group of Poles in the corner said crossly: "Half-wit!" but it was not clear whether this referred to the warder or to Julia.

Julia, deaf to everything, tore off her clothes, lay down, and pulled her blanket over her head, ostentatiously demonstrating her detachment from her neighbors, in every one of whom she, as an orthodox Stalinist, scented the "true enemy."

The rest of us, shaken by this incident, went to bed quickly. Next to me was the Latvian Milda, an elderly woman who had all the marks of an impeccable member of the working class. She looked like a washer-woman, with deep-set eyes, a flat chest and big belly, long thin arms and large hands with prominent veins. The charge against her was that she had lived it up with foreigners in elegant restaurants, seduced Soviet diplomats, and wormed secret information out of them. This was July 1937, when no one cared any longer whether the charges bore the slightest semblance of probability.

Before lying down, Milda combed her thin yellow hair neatly and drew from under the straw pillow some cotton wool with which she carefully stopped both her ears. She offered me some and, when I looked surprised, explained:

"It was winter when I was arrested, so I have my overcoat, and I pull this out of the lining."

"But what's the idea of stopping your ears?"

She shrugged wearily.

"So as not to hear. So as to get some sleep."

But I did not stop up mine. Rather than play the ostrich, I would see things through to the finish. And I saw them through on that hot July night of the year 1937.

It began all at once, not by degrees or with any sort of prelude. Not one, but a multitude of screams and groans from tortured human beings burst simultaneously through the open windows of our cell. In the Butyrki prison, an entire wing on a certain floor was set aside for night interrogations, and it was doubtless equipped with the latest refinements of the torture chamber. Klara, the ex-victim of the Gestapo, assured us that the implements used here must have been imported from Hitler's Germany.

Over and through the screams of the tortured, we could hear the shouts and curses of the torturers. Added to the cacophony was the noise of chairs being hurled about, fists banging on tables, and some other unidentifiable sound which froze one's blood.

Although these were only sounds, they conjured up such a vivid picture that I felt I could see it in every detail.

I imagined all the interrogators as looking like Tsarevsky. As for the victims, I could see them before me with that unmistakable look of theirs—no, I cannot describe it, but to this day I can tell former "guests" of the NKVD by the trace of that look which flickers somewhere deep down in their pupils. To this day, in the 1960's, I can startle people I meet in the train or at the seaside by the clairvoyant question: "You were in, weren't you? You've been rehabilitated?"

How long could it go on? Till three o'clock, they had told me. But surely no one could endure this longer than a minute. Yet the noise went on and on, dying down from time to time and then bursting out again. An hour—two, three, and four hours, from eleven till three o'clock every night. I sat up in bed. An old Eastern saying came into my mind: "May I never experience all that it is possible to get used to." Yes, my cellmates had got used even to this. Most of them were asleep or, at all events, were lying quietly, their heads buried under blankets in spite of the stifling heat. Only a few newcomers, like me, were sitting up in their bunks. Some had stuffed their fingers into their ears, some were as if petrified. Every now and then the supervisor would stick her head through the flap-window with the command:

"Lie down! It's against the rules to sit up after lights out."

Suddenly a long-drawn-out shriek of pain was heard, not from a distance but in our cell. A young woman with a long, disheveled braid rushed to the window and, as if demented, beat against it with her head and hands.

"It's him! I know it is, it's his voice! I don't want to go on living, I don't, I can't! Please let them kill me. . . ."

Several women jumped up and surrounded her, dragging her away from the window and trying to convince her that she was mistaken, that it was not her husband's voice. But she was not to be comforted. No, she would know his voice among a thousand—it was he they were torturing and mutilating; and was she to lie here and be silent? If she screamed and made a disturbance they might kill her right away, and that was all she wanted—how could she possibly live after this? . . .

We heard steps in the corridor: the door was flung open and the supervisor came in, accompanied by a senior warder. The latter, with a practiced movement, pinned the woman's arms behind her back and forcibly poured down her throat some liquid from a glass, saying: "Drink this, it's valerian drops."

I doubt, though, that any valerian would have had the effect of making the woman collapse almost instantaneously onto her bunk, close her eyes, and fall into a strange, deathlike sleep.

The cell was quiet again. Milda raised her head, felt under her pillow, and again offered me some cotton wool.

"I don't want it, thanks. Tell me, who is that woman?"

"Oh, she's one of the Poles. There are seven of them in that corner. Her husband's a Russian, a Soviet citizen. They married a short time ago, and there's a baby, three months old. They had to bandage her breast to stop the milk. What she can't stand is the thought that her husband was arrested because of her, for associating with a foreigner."

By now it was nearly three o'clock, and the sounds were dying down. Once more I heard a chair thrown to the ground, once more a man's stifled sob. Then silence.

I could see in my mind's eye the torn, bloody victims as they staggered or were carried out of the torture chamber.

I could see the interrogators putting their papers away till the following night.

"Give me some cotton wool now," I begged Milda.

"You won't need it now. It's all over till tomorrow."

"Never mind, please give me some."

She shrugged in bewilderment, but gave me a piece of the gray material. I stopped both my ears, then pulled over my head the prison blanket, smelling of dust and grief, and took a corner of the straw pillow firmly between my teeth. It would be easier like this: I could neither hear nor see. If only I could stop thinking as well! . . .

I knew I should not sleep until I had repeated some poem over to myself ten or a hundred times. I chose Michelangelo's lines:

Sweet is't to sleep, sweeter to be a stone.
In this dread age of terror and of shame,
Thrice blest is he who neither sees nor feels.
Leave me then here, and trouble not my rest.

28

• *In accordance with the law of December 1st*

In the Butyrki prison, isolation from the outside world was much more absolute than at Kazan. The cells were filled on the principle of grouping together people at the same stage of the interrogation process. Consequently no one joined us direct from outside: any newcomers were either, like me, at the end of their interrogation or at any rate close to it.

We suffered a great deal from our enforced ignorance, but for the rest we managed to achieve some sort of daily routine. The dreadful nights alternated with busy days. Indeed, we had scarcely a moment's free time from reveille to lights out. There were the emptying of the enormous slop bucket, the long queues for the washroom, the distribution of food three times a day from large pails, washing up, mending our frayed stockings and underwear—none of us were allowed to receive parcels—the daily exercise, the taking of orders for the "store" from those who were lucky enough to have a little money, the exchange of books, inspection and roll call—all this filled and even over-filled the daylight hours. By day our cell resembled the hold of a disabled ship, tossing about at the mercy of winds and waves. Just as on a sinking ship, the reactions of individuals varied from the elaborately calm to the exalted and the cowardly. Though of these last there were few.

A day or two after my arrival there was a row over our feeding the birds. It came to the ears of Popov, the prison governor, that we were scattering crumbs out of the window every evening and that the sparrows, having discovered this, would crowd about our window, making a great din; they would even get into the cell and fill it with their twittering, to the delight and entertainment of us prisoners.

Popov, with an escort of warders, burst into the cell at an unscheduled hour and, in a voice choked with fury, made a short but trenchant speech on the theme that this was not a pleasure resort. Every sentence ended with the refrain: "Don't you forget that you're in prison, and in the Butyrki prison at that!" However, we were not put in punishment cells or deprived of our books or exercise. It was said, in fact, that Popov was not such a bad fellow as all that, and that his bark was worse than his bite. In any case, he was soon to have a chance to appreciate his own formula, "the Butyrki prison at that." Before two or three months were out, he had ceased to be the governor of the jail and become one of its inmates.

From time to time one of us was called out of the cell. If we were told to "bring our things," the others would turn pale and whisper with dried lips, "She's to go before the court" or "She's to be told her sentence." We knew, indeed, that some people were sentenced by the courts and others, *in absentia*, by a "special board" of the NKVD; but we had no idea as yet what the sentences were. Much argument raged about this: some spoke chillingly of ten years or even death, but most of us brushed them aside angrily. We relied on the comforting thought: "If Zinovyev and Kamenev, Pyatakov and Radek got only ten years each, then surely we small fry . . ."

If one of us was summoned "without her things," the cell would be thrown into a flutter for a different reason. As soon as the door had shut and the key turned, ominous whispers would be heard in every corner:

"Now I wonder what that's for. Her interrogation was over a long time ago."

"How dare you! She's a decent person."

"I always thought so, but . . ."

"And I, like a fool, talked my head off to her last night."

It was like a psychosis: good and sensible people who had been on friendly terms would suddenly see in their neighbors potential spies and *provocateurs*. Afterward they would often feel ashamed of these fits of distrust and suspicion, this sense of being a wolf among wolves. But a few hours later someone else would be called out "without her things," and the rest would again be petrified with fear. Suppose she were to repeat to the interrogator everything that had been said in the cell the day before?

So when, one bright day that summer, the supervisor put her head through the flap-window and quietly pronounced my name, my first feeling was acute embarrassment. Why without my things? What on earth would my cellmates think of me? . . .

It is curious how a prisoner's traumatic state can affect his reflexes. This was my first call after three weeks in Butyrki, and one would have supposed that my first thought would be of my trial, the sentence that awaited me, whether I was to live or die. But no—I was only worried about whether my cellmates would think I had been called out to inform on them. . . .

I followed the warder's whispered directions almost mechanically through the maze of corridors, till I suddenly found myself back in the central "reception hall" of the prison.

"Over here!" The bolt shot to, and I was again inside one of the "kennels"—a tiled booth with standing room only. Was I to be taken away somewhere?

Leaning against the cold wall, I once more lost track of time. The tiles glittered in the light of a strong bulb. It was still there when I closed my eyes, only less bright. Well, they could not leave me here for ever.

The bolt clicked open. A young officer stood in the doorway and thrust a piece of paper at me.

"Read this." Before I could ask him anything, he had locked me in again.

"This" was my charge sheet, signed by Vyshinsky.* So

---

* Andrey Vyshinsky, 1883-1954, notorious Soviet jurist who acted as prosecutor in the show trials of 1937-38. Later Foreign Minister and Soviet representative at the United Nations.

he had endorsed it. . . . I remembered meeting him once at a summer resort. He wore an embroidered Ukrainian shirt and was accompanied by his ailing, emaciated wife and their daughter Zina, with whom I used to go bathing every day. Had he remembered me when he signed this paper? Or were all names confused in the bloody haze of his mind? He had, after all, not shrunk from sending to execution his old comrade Yevgeni Veger, the secretary of the Odessa regional committee; so why should he spare his daughter's holiday friend?

I ran my eyes over the "preamble" to the act of accusation. There was nothing special here—the usual newspaper claptrap: "A Trotskyist terrorist counter-revolutionary group . . . dedicated to the restoration of capitalism and the physical annihilation of Party and government leaders." These formulae, repeated millions of times over, had lost their power to shock and now inspired only a vague nausea: they affected the mind almost like a refrain, or like some oft-repeated fairy tale. People skipped them and would wait with bated breath for the real story to begin, for the ogre to appear. . . .

In my case, after the ritual introduction came a list of "members of the counter-revolutionary Trotskyist terrorist organization among the editorial staff of the newspaper *Red Tartary.*" Again not the slightest attempt at plausibility. Some of the people on the list had never worked on the newspaper; others had long since moved to other towns and were nowhere near at the time of the crime. As I later found out, those who had moved away far enough were never arrested. . . .

I read on. Ah, here was the real story, here came the ogre at last. "On the basis of these facts, the case is referred to the military tribunal . . . in accordance with sections 8

and 11 of Article 58, of the Criminal Code and the law
of December 1, 1934."

By now the flutter in my temples had changed to a slow,
reverberating throb. What law was this? Its date boded
no good. . . .

The young officer opened the door of my kennel again.
This time I was able to take in his appearance: he had a
sharp little nose and a toothbrush mustache, like the comic
policeman in Gorky's play *Enemies*. . . . As if from very
far off, I heard him asking for the second time: "Have you
acquainted yourself with the indictment? Is it quite clear?"

"No, it isn't. I don't know what the law of December 1,
1934, says."

He looked astonished, as if I had asked him what the
sun or the moon were. Then he replied with a shrug:

"It says that the sentence must be carried out within
twenty-four hours of its being pronounced."

Twenty-four hours—plus another twenty-four between
now and my trial. (They had told me in the cell that
people were usually tried on the day after they were shown
the indictment.) That made forty-eight hours in all—forty-
eight hours to live.

There was once a little girl called Genia. Her mother
used to braid her hair, and she grew up and fell in love
and tried to discover what life was about. And she lived
as a grown-up woman for two whole years, till she
was twenty-eight. And she had two sons, Alyosha and
Vasya. . . .

The cell was dead silent. It was the first case of its kind
known to us. No one from our cell had been before the
military tribunal: only before various summary or civil
courts. And no one had yet been presented with an indictment and the clause about the sentence being carried out

within twenty-four hours. No one had the slightest doubt what was in store for me tomorrow. They stroked my hair, took my shoes off, made me swallow some veronal which had by some miracle been smuggled past the guards: but it had no effect. The organism refuses to waste the last hours of its existence on sleep.

All that night, unreproved by the supervisor, I sat at the table in the middle of the cell. I learned then of what kindness my fellow inmates were capable. It was hard to believe they were the same women who had suspected one another of the blackest treachery. They learned by heart my children's names and my relatives' addresses, so that if they themselves survived they could tell them of my last hours.

One would need to be a Tolstoy to give an account of the thoughts and feelings of a person condemned to death. When I think of that night, I can only remember a curious sharpness in the outlines of objects and a painful dryness in my mouth. As for my thoughts, an exact record of how they ran would read strangely. "Do people have time to feel pain when they are shot? What on earth will Alyosha and Vasya say when they have to fill in forms about their parents? What a shame about that new silk dress that fitted me so well and that I never had a chance to wear. . . ." Such, more or less, were the thoughts that went through my head.

Some books were lying on the table. I opened one. Baransky's *Economic Geography*. Good. I could take another look at a map. There was the world. And there was Moscow, where I had been born, and where now I was about to die. There were Kazan, Sochi, and the Crimea. And there was the rest of the world, which I had not seen and never would.

At dawn some sparrows, who had evidently not heard that this was not a pleasure resort and that Popov, the prison governor, had categorically forbidden all contact between birds and prisoners, perched boldly on the edge of the wooden screen, their tails fluttering comically. With joyful voices they ushered in the grandest month of the year. It was the morning of August 1, 1937.

29

• *"A fair and speedy trial"*

In the Lefortovo prison, all the doors open noiselessly. Soft carpets drown the sound of footsteps, and the warders are exquisitely polite. The "kennels" have stools on which one can sit, and their tiled walls are so dazzlingly white that they remind one of an operating theater.

The solitary cell in which I was put that morning of the first of August was as clean as a private room in a hospital, and the wardress resembled the matron of a holiday home. Here I was to await my trial. I remembered Garey's rule of thumb: "The cleaner and more polite, the nearer to death."

Nevertheless, the atmosphere of the place made me want to smarten up my own appearance. I got out of my bundle my blue "afternoon" dress, spent a long time straightening its folds, used my fingers as curlers, and put tooth powder on my nose. All this I did more or less mechanically. There was nothing strange about it: Charlotte Corday made herself look pretty before she was guillotined, and so did Madame Camille Desmoulins, not to mention Mary Queen of Scots. But all these thoughts about my appearance were quite self-contained and had no effect on the great black toad which sprawled close to my heart, and which nothing would drive away.

My hour had come. The military tribunal of the Supreme Court—three officers and a secretary—faced me across a table. On either side of me were two warders. Such were the "fully public" * conditions in which my "trial" was conducted.

I looked intently at my judges' faces, and was struck by the fact that they closely resembled one another and also for some reason reminded me of the official at Black Lake who had taken my watch. They all looked alike, although one of them was dark and another's hair was graying. Yes, I saw what it was they had in common— the empty look of a mummy, or a fish in aspic. Nor was this surprising. How could one carry out such duties day after day without cutting oneself off from one's fellow men—if only by that glazed expression?

Suddenly I found that it was easy to breathe. A summer breeze of extraordinary freshness was blowing through the wide-open window. The room was a handsome one, with a high ceiling. So there were, after all, such places in the world!

Outside the windows there were some large, dark-green trees: their leaves were rustling and I was moved by the

* According to Soviet law everybody has the right to public trial.

cool, mysterious sound. I could not remember having
heard it before. It was strangely touching—why had I
never noticed it?

And the clock on the wall—large and round, its shiny
hands like a gray mustache—how long was it since I had
seen anything like it?

I checked the time at the beginning and end of the trial.
Seven minutes. That, neither more nor less, was the time
it took to enact this tragicomedy.

The voice of the president of the court—Dmitriev,
People's Commissar for Justice of the Russian Republic—
resembled the expression of his eyes. If a frozen codfish
could talk, that is undoubtedly what it would sound like.
There was not a trace here of the animation, the zest
which my interrogators had put into their performances:
the judges were merely functionaries earning their pay.
No doubt they had a quota, and were anxious to over-
fulfill it if they could.

"You have read the indictment?" said the president in
tones of unutterable boredom. "You plead guilty? No?
But the evidence shows . . ."

He leafed through the bulky file and muttered through
his teeth: "For instance the witness Kozlov . . ."

"Kozlova. It's a woman—and, I may add, a despicable
one."

"Kozlova, yes. Or again, the witness Dyachenko . . ."

"Dyakonov."

"Yes. Well, they both state . . ."

But what they stated the judge was too pressed for time
to read. Breaking off, he asked me:

"Any questions you wish to ask the court?"

"Yes, I do. I am accused under section 8 of Article 58,

which means that I am charged with terrorism. Will you please tell me the name of the political leader against whose life I am supposed to have plotted?"

The judges were silent for a while, taken aback by this preposterous question. They looked reproachfully at the inquisitive woman who was holding up their work. Then the one with grizzled hair muttered:

"You know, don't you, that Comrade Kirov was murdered in Leningrad?"

"Yes, but I didn't kill him, it was someone called Niko-layev. Anyway, I've never been in Leningrad. Isn't that what you call an alibi?"

"Are you a lawyer by any chance?" said the gray-haired man crossly.

"No, I'm a teacher."

"You won't get anywhere by quibbling. You may never have been in Leningrad, but it was your accomplices who killed him, and that makes you morally and criminally responsible."

"The court will withdraw for consultation," grunted the president. Whereupon they all stood up, lazily stretching their limbs. . . .

I looked at the clock again. They couldn't even have had time for a smoke! In less than two minutes the worshipful assembly was back in session. The president had in his hand a large sheet of paper of excellent quality, covered with typescript in close spacing. The text, which must have taken at least twenty minutes to copy, was my sentence: the official document setting forth my crimes and the penalty for them. It began with the solemn words: "In the name of the Union of Soviet Socialist Republics," followed by something long and unintelligible. Oh, yes, it was the same preamble as in the indictment, with its "res-

toration of capitalism," "underground terrorist organization," and all the rest. Only wherever that document had
said "accused," this one said "convicted." . . .

The president went on in a whining nasal voice. How
slowly he read! He turned over the page. Here it comes
. . . any moment now he would utter the words "supreme
penalty . . ."

Again I heard the trees rustling. For a moment I felt as
if all this were part of a film. How could they possibly
be going to kill me for no reason at all—me, Mother's
Genia, Alyosha's and Vasya's mother. By what right . . .

I thought I must have screamed, but no—I stood and
listened in silence, and all these other strange things were
taking place inside me.

A sort of darkness came over my mind; the judge's
voice sounded like a muddy torrent by which I might be
swamped at any moment. In my delirium I noticed a
thoughtful move by the two warders who were joining
hands behind my back—evidently so that I should not hit
the floor when I fainted. Why were they so sure I would
faint? Oh, well, they must know from experience, doubtless many women fainted when they heard their death
sentence read out.

The darkness closed in again—for the last time, surely.
And then, all of a sudden—

What was it? What had he said? Like a blinding zigzag
of lightning cutting across my mind. He had said . . .
Had I heard right? . . .

"Ten years' imprisonment in solitary confinement with
loss of civil rights for five years . . ."

The world around me was suddenly warm and bright.
Ten years! That meant I was to live!

". . . and confiscation of all personal property."

To live! Without property, but what was that to me?
Let them confiscate it—they were brigands anyway, con-
fiscating was their business. They wouldn't get much good
out of mine, a few books and clothes—why, we didn't even
have a radio. My husband was a loyal Communist of the
old stamp, not the kind who had to have a Buick or a
Mercedes . . . Ten years! . . . Do you, with your codfish faces, really think you can go on robbing and murder-
ing for another ten years, that there aren't people in the
Party who will stop you sooner or later? I knew there
were—and in order to see that day, I must live! In prison,
if needs be, but I must at all costs live!

If the "judges" ever looked at their victims' faces, they
could not have failed to read all these silent exclamations
in mine. But they did not look at me. Having read the
sentence, they rose and hurried away. In single file they
trooped out of the room. This time, no doubt, they would
have time for a smoke, and then back to the grindstone.
A hard day's work . . .

I looked around at the guards, whose hands were still
clasped behind my back. Every nerve in my body was
quivering with the joy of being alive. What nice faces
the guards had! Peasant boys from Ryazan or Kursk, most
likely. They couldn't help being warders—no doubt they
were conscripts. And they had joined hands to save me
from falling. But they needn't have—I wasn't going to fall.

I shook back the hair curled so carefully before facing
the court, so as not to disgrace the memory of Charlotte
Corday. Then I gave the guards a friendly smile. They
looked at me in astonishment.

30

• *Penal servitude—what bliss!*

"I don't suppose you'll want dinner?" the wardress, who looked like a matron, asked me. She, too, knew what to expect: people had no appetite after being sentenced.

"Not want dinner? Yes, of course I do!" I replied cheerfully, and set about briskly repacking my bundle while I waited for the food. I had heard that, unless they were sentenced to death, prisoners were not kept at the Lefortovo prison but sent back to Butyrki.

I looked forward to the move. It meant a common cell, people, companions in misfortune.

Dinner arrived. Not in mess tins but in enameled bowls. Meat soup and *kasha* with butter. A good meal, for people condemned to death, of whom there were so many in this prison. It was given, perhaps, as a humane gesture in keeping with the tradition going back to the "rotten liberal" Nicholas II.

I ate everything down to the last scrap, and made a resolution from now on to eat, sleep, and do exercises every morning. I intended to stay alive. Just to spite them. I was consumed by the desire to survive the tragedy which had befallen our Party. More than ever I felt sure that they could not destroy it completely, that there were people in it who would stop them. Keep alive . . . Keep alive . . . Grit your teeth . . . Grit your teeth . . .

As I repeated these words to myself, I was reminded of Pasternak's lines in his poem "Lieutenant Schmidt":

The indictment stretched, mile on mile,
Pit-shafts mark our weary way.
We greet our sentence with a smile—
It's penal servitude! What bliss!

Suddenly these words thrilled me with their aptness. It is only at such times that one realizes the true value of poetry, and one's heart fills with tender gratitude toward the writer. How could Pasternak have known so exactly what one felt, living in his "melancholy Moscow home"? I remembered other lines: "The rest were drunk with space, and spring, and penal servitude. . . ."

If only he could know how much his poem helped me to endure, and to make sense of prison, of my sentence, of the murderers with frozen codfish eyes.

It was getting dark. Here, too, the window was covered with a wooden screen as well as bars. For some reason, they were late switching on the light. I could not wait to be back in Butyrki, away from this place where death stared at you from every corner. I rested my head on the table and mentally recited "Lieutenant Schmidt" from beginning to end. I was terribly moved by the lines:

The wind, warm and selfless, caressed the stars
With something of its own, eternal and creative . . .

I was awakened by the ritual command:
"Take your things!"
By now it was quite dark. Above the screen I could see the stars, the same stars as in Pasternak's poem, but the light had still not been switched on. From every corner, from the walls painted dark-red, terror closed in on me. "Take your things!" Yes, quickly, yes. . . . Back to Butyrki! It now seemed to me like home, and I imagined the consolation of being in a large cell full of people who

sympathized with me and whom I knew. Let the cell be like a sinking ship—there was still a chance that it might stay afloat. Here, there was not. It was the seventh circle of Dante's hell, where nothing lay ahead but death, and I could not wait to get out of it.

For a second I panicked. Suppose it was not to Butyrki they were taking me, but to the cellars? The notorious Lefortovo cellars where, according to the stories whispered at Butyrki, they used tractor engines to drown the sound of shooting . . . There too, no doubt, the walls were painted dark-red, so that the blood would not show. . . .

It took an unbelievable effort to pull myself together. What nonsense! Had I not heard the sentence with my own ears—ten years in a security prison. . . .

"Ready?"

"Yes, yes."

I was led down a long corridor, past a row of single cells. Doors, doors . . . and down flights of steps. My heart missed a beat. Could it be the cellars after all? No! A gust of pure night air struck my face. A courtyard, and a Black Maria . . .

Once again I was locked into a box smelling of oil paint, in which it was possible to sit but not to stand. The engine started. We were off, "home," to Butyrki. Death, which had crouched at my back for two days, slunk off, disappointed.

As I recovered from my mortal terror, I lost control over myself. It was useless to repeat Pasternak's consoling lines about the bliss of penal servitude. Nothing helped. A suffocating lump rose in my throat and I broke into wild, uncontrollable sobs. I was bursting with indignation. "You swine, how can you do this to human beings, and

Communists at that!" I must have been shouting this aloud. I was quite beside myself and battered the locked door with my fists and beat my head against it.

The soldier who opened the door was the living image of the one from Pskov, in the film *The Kronstadt Boys.* He had the same good-natured, rather stupid face, with raised, arched, colorless eyebrows. The words he used to calm me down at once dispelled the atmosphere of horror and brought things down to the level of a village quarrel.

"Steady now, my girl! If you cry like that your face'll swell up, and none of the boys'll look at you!"

I was glad he had talked to me in this familiar way. It meant we were out of the Lefortovo zone of lethal politeness. I had an almost physical sense of his kindness, of his unsophisticated human heart. I sobbed louder and more desperately than before, but now in the secret hope that he would continue to comfort me.

"I'm not a girl. . . . I'm a mother, I've got children. Can you understand it, comrade, I've done nothing, absolutely nothing wrong at all. . . . And they . . . Do you believe me?"

"Of course I believe you!" He looked surprised. "If you'd done anything wrong, would I be taking you back to Butyrki now? You'd have stayed at Lefortovo, but stop shouting, do! Listen, I'll leave the door open, so you'll get a breath of fresh air. Would you like some of them laverian drops? We've got some, if you want any. . . . Go on, have a good breath of fresh air. . . . There's no one else in the van. . . . I had to make a special trip just for you. They'd forgotten you or something, and now here you are riding all alone through Moscow, like the Queen of Sheba. . . ."

"Ten years! Ten years! What for? How dare they, the swine!"

"I've got a chatterbox on my hands, that's sure! Shut up, will you! Of course you're not guilty! Would they have given you ten years if you had been? Do you know how many a day they're finishing off now? Seventy! That's how many. . . . I reckon it's only women they're letting off. . . . You're the third I've brought back in the last day or two."

I was instantly silenced by this statistical revelation. Evidently, too, the scale of the operation was such that they no longer had time to brief the guards—God help this one if they came to hear of what he had just told me. . . . But I would keep my mouth shut.

"Well, you going to be a good girl now? That's right. The way you were yelling at me, I might have been your husband. . . ."

I drank the "laverian" drops he gave me, and at once, lulled by the steady chug of the engine, felt an overpowering desire to sleep. I was dropping off when I heard the soldier's comforting whisper:

"You'll never stay ten years out there. Not on your life! Maybe a year or two at the worst. Then you'll invent some gadget or other and they'll let you out, to go home to your children."

In his kindly, muddled mind, the horrors of the present coexisted with old tales of scientists released early as a reward for making an invention. But I so much wanted to believe him. And how nice it was to rattle along in a Black Maria, when the door of the cage was open and the guard was a boy from Pskov who didn't care if he broke the rules. And soon we would be at Butyrki. Penal servitude —what bliss!

31

• *The Pugachev Tower*

The "special block," with its cleanliness and its folding beds, was no longer for me. I was now in transit, a deportee, and must be lodged accordingly. I was put in the Pugachev Tower, which had actually been used to imprison the rebels against Catherine the Great. Anna Zhilinskaya, a historian, with whom I shared a plank bed, explained to me the architecture of the building with its slits of windows and winding stairs.

By "shared" I mean that we slept on the same plank bed, taking turns. The beds were a single bench all around the walls. The cell was twice as full as it should have been, and those for whom there was no room on the bench slept on the stone floor or on the large unpainted table in the middle of the cell.

That year, August was stiflingly hot in Moscow. Once again, as in Kazan, we sat in nothing but our pants and brassières, dirty and sweating. There were new arrivals every day and no room to put them in. But the prison authorities were not in the least worried—after all, these were only prisoners awaiting deportation.

No one was allowed parcels, nor were orders taken for the "store." We had nothing but our rations.

My cellmates represented a much broader section of the population than they had in the special block. There were many peasants, factory workers, shop girls. They were mostly "babblers," "people whose tongues were too long," that is "anti-Soviet propagandists," nearly all of them sentenced to five or eight years under Article 58, section 10.

My own sentence—ten years in prison, in solitary confinement, and imposed by a military tribunal at that—caused something of a sensation in the cell.

This was still, of course, before October 1937, when twenty-five-year terms were introduced. At the time, ten was the maximum, short of death, and conferred a certain aura of martyrdom.

People with my kind of sentence were commonly thought to belong to the highest ranks of Soviet society. Thus a rumor spread that I was Pyatakov's wife, and I found it hard to convince my cellmates that I was not.

The only other one of us who had been sentenced to so long a term as mine was "Grandma Nastya," a peasant woman of sixty-five from a collective farm near Moscow. Why she had drawn such an unlucky number in the lottery it was hard to say. Nobody could make anything of the accusation of "Trotskyist terrorism" against Grandma, with her mild, wrinkled old peasant woman's face and her sad, childlike eyes. Grandma herself was more puzzled than anyone else and, when she heard that I was in the same boat, brought her bundle over to my place on the plank bed and sat at my feet. The small bundle of clothes was in an ancient canvas bag that evoked the village Russia of bygone days. As Grandma looked at me with her trustful eyes, I felt the same burning shame as I had felt among the German Communists in the Comintern cell.

"Are you one of those tractorists too, dearie?"

"No, Grandma. I'm just an ordinary woman. A teacher. A married woman with children. All this nonsense is something the interrogators and the judges have thought up, it must be they who are the traitors. What you and I must do is hold on—it'll all come out in the end."

Grandma nodded, her head swathed to the eyebrows in a kerchief.

"Yes, likely enough. . . . They've invented it about me too, they put me down as one of those tractorists. But as God's my witness I never went near one of them cursed things. I don't know how they thought of it. . . . They don't put old women like me on tractors. . . ."

Someone near us burst out laughing, and Anna Zhilinskaya muttered sleepily:

"You'll be the death of me, Grandma, that's the best thing you've said yet!"

But I couldn't laugh. I was ashamed. When would I at last stop feeling ashamed and responsible for all this? After all, I was the anvil, not the hammer. But might I too have become a hammer?

Since my trial and sentence, I had found that tears came easily, and as I looked at Grandma Nastya now, I felt a lump in my throat. My mother was eight years younger than this woman, but it was unbearable to think of her in the same situation.

I didn't know at the time that my parents had been arrested. They were released two months later, but my father died soon afterward and my mother fell ill with diabetes, of which she too eventually died.

Deportees were allowed no books, so there was even more conversation here than in the special block. Each of us told story after story . . . not only about her past life but about all she had seen and heard in prison. Apart from this, we were all very concerned with geography, the names which most often rose above the general hubbub

being Kolyma, Kamchatka, Pechora, Solovki.* But none of
these concerned me—not for me the bliss of penal servitude. What lay ahead of me was solitary confinement.
Some of my cellmates, better informed than I, mentioned the places in which I should be interested—Suzdal,
Verkhneuralsk, Yaroslavl, where there were security
prisons for those held in solitary.

Anna Zhilinskaya tried to comfort me. She had heard
these places were not so bad. You were given books. They
didn't starve you. But these illusions were soon dispelled.
One day at dawn another group of deportees were
squeezed into our cell.

"Never mind, make room! A large convoy will be going
off soon," the wardress snapped in answer to our protest.

Among the new arrivals was a Moscow Party worker
named Raissa (I don't remember her surname), who knew
for a fact that the Central Committee had had a plenum
in July at which the "boss" had made a speech complaining
about prison discipline. Places of detention, he maintained,
had been "turned into health resorts," and political prisons
were the most lax of all. It was easy to imagine with what
fervor the authorities would set about tightening the
screw. Would we survive it? There were three of us interested in this point—Anna Zhilinskaya, Tanya Andreyeva
from Harbin, and myself.

Tanya reminded me of Lyama. She was equally practical, kind, and full of sympathy for her companions in
misfortune. I listened with interest to her tales of Shanghai,
where she had lived for a long time, of the Russian émigré
community, of her arrival in the Soviet Union, where she

---

* Kolyma and Kamchatka in the Soviet Far East, Pechora and Solovki
in the White Sea region, notorious concentration camp areas.

had come to join her husband, a Communist, and of their arrest.

Tanya had been sentenced to eight years in camp, but she was full of optimism.

"I'll come through, I'm certain! I'll do manicures and pedicures for the wardresses, and show them foreign hair styles." She laughed, screwing up her narrow eyes, which, as she said, had "gone Chinese." "Besides, I've got a lot of odds and ends of silk I can use to bribe them with so they won't be too hard on me. Look!"

She undid her bundle and amid the stench and heat of the Pugachev Tower, some Chinese silk jackets magically burst into blossom.

By contrast, Anna was full of fear and gloomy prophecies.

"They won't help you, Tanya—they'll all get stolen on the way. They'll do us all in, you can be sure of that. It's just a question of how long it will take. You two weren't in the Lubyanka, you don't know what it's like. I had three months of it."

"Well, I was at Lefortovo," I boasted mildly.

"That's the last scene of the play, where all the characters get shot. All except a few lucky ones like you. Lubyanka's something else again. It's the worst stage of the investigation. You should have seen my interrogator—a woman. A monster! A real hyena . . ."

One night Anna told me about her cellmate at the Lubyanka, a Party member called Eugenia Podolskaya.

"Look, Genia, I feel it in my bones I won't get out of this alive. So I must give the message to someone. I promised Eugenia I'd let her daughter know."

"Did she die?"

"I'm sure she did. But are you willing to let me tell you?

Yezhov said he'd have anyone who knew about it shot. . . ."

For greater privacy, we went to the washroom. This was not at the end of the corridor as in other prisons, it was a small annex to the cell. We stood by the narrow tall window with its elaborately intertwined bars which evoked memories of the eighteenth century, Pugachev's rebellion, and the headsman's ax. It was there that Anna, talking rapidly, her eyes dilated and shining, told me the story.

One night in the Lubyanka cell she shared with Eugenia, Anna woke up to a faint sound like that of dripping water. It was blood pouring from her cellmate's wrist. She had cut it with a small razor blade she had somehow managed to steal from an interrogator. Anna shouted for help, and the warders came and took Eugenia away. When she returned three days later, she told Anna that she was determined to kill herself one way or another. It was then that Anna promised to tell her daughter everything if she herself survived.

When Eugenia had first been summoned by the NKVD, she had felt no alarm. She thought, as an old Communist, that they wanted to entrust her with some important mission. So they did. The interrogator asked her, to begin with, whether she was ready to take on a difficult and dangerous task for the Party. She was? Very well. She would have to spend a short time in a cell. It wouldn't be for long. Once she had carried out her mission, she would receive new papers and a new name, and would have to leave Moscow for a while.

The mission was to sign a number of statements about the subversive doings of a certain counter-revolutionary group, to which, for the sake of plausibility, she would confess that she belonged.

"In fact, give evidence of something she knew nothing whatsoever about?"

"Well, she had the NKVD's word for it, hadn't she? They said they knew for certain that the group had committed the most monstrous crimes. All they needed her signature for was to give the case a certain legal weight. Besides, there were top-level considerations of the sort an ordinary Party member didn't need to know if she were really willing to perform a dangerous task."

Step by step, she followed him into this maze of duplicity. Finally, they stuck a pen into her hand and she began to sign. She spent her days in the common cell, and at night they called her upstairs, gave her a good meal, and let her sleep on the sofa. Then one evening, when she came up, she found a different officer, who looked at her ironically and said:

"And now, my dear, you're going to be shot."

In a few simple words he explained to her the role she had played as a stool pigeon, added some coarse abuse, and told her that her depositions had enabled them to "write off" a group of no less than twenty-five people. After that, she was sent back to the cell and not called out for over a month. That was when she used the razor blade she had somehow managed to steal in the interrogator's office.

"She was one of those," said Anna, "who out of pure fanaticism, without a thought for their own advantage or safety, destroyed many others and themselves as well. Her remorse was so terrible that I myself came to believe it was better for her to die. I gave up trying to dissuade her and promised, as she asked me, to tell her daughter everything if I survived."

And now Anna was confiding the secret to me, although my sentence was longer than hers. I promised to pass it on, and memorized the address of Eugenia's daughter. Unfortunately, I was never able to keep my promise, because by 1955, when I arrived in Moscow after eighteen years' absence, I found I had completely forgotten the daughter's address and even her first name. My memory, good though it was, had been burdened by too many things.

Perhaps it was as well. What good could it possibly have done the daughter to learn of her mother's tragedy, which had caused the death of so many people?

I spent only a fortnight in the Pugachev Tower, but it was a trying time. The worst part were the nights when it was Anna's turn to sleep, and I sat on the edge of the bed at her feet, half awake and struggling with nightmares.

At such times, the horrible medley of human beings moaning and groaning and scratching themselves seemed a huge mass grave in which people who had been shot lay waiting to die.

One such night, the block supervisor and three wardresses appeared with long sheets of paper in their hands. The big convoy was leaving. The list of names was read out. People jumped up and feverishly grabbed their rags, as if in them lay their one hope of salvation. The big convoy . . . The big convoy . . .

32

• *The Stolypin coach*

The railroad coaches had not been renamed. They were still called after Nicholas II's Prime Minister Stolypin,* and no one seemed surprised at this. They were gloomy but clean. I was to appreciate them fully only two years later, when I spent over a month traveling from Yaroslavl to Vladivostok in a wooden freight train. At the moment, we were mortally depressed by the massive gratings and the reinforced guard.

We were, of course, taken to the train in a Black Maria, and boarded it at some out-of-the-way siding, but I had caught a distant glimpse of the Northern station and knew this meant that we were going to Yaroslavl. It was the worst of our three possible destinations. I had heard a lot in the Pugachev Tower about the special wing of the Yaroslavl prison, built after the 1905 revolution for especially important political prisoners. Its old tradition of exceptional severity had, by all accounts, survived to our day. I had so much hoped for Suzdal. There the prison was housed in a former convent, and I had often comforted myself with the thought that a monastic cell must be more humane than a prison cell. Verkhneuralsk was also said to be much better than Yaroslavl.

"Ach, Genossin, wir sind doch bekannt!"

I immediately recognized the German film star Carola

---

* Stolypin was associated with the deportations after the 1905 revolution.

Heintschke, the one who had hidden her gold rings in her hair during that first memorable search at Butyrki. She had changed greatly since then. The gold of her hair had lost its luster, and pathetic little wrinkles showed at the corners of the mouth. Yet she was even more fascinating than before. Her face was ivory-white, without the faintest trace of color. She had a childlike smile and sad, amber-colored eyes.

Her sentence was the same as mine, but it was a thousand times worse for her because, to add to everything, she knew no Russian, and no one in her cell spoke German.

Now, remembering the few chance words we had exchanged at our first meeting, she was overjoyed to find someone who could speak her language, however poorly.

She had heard nothing about her husband but felt sure that he was no longer alive. There was no mistaking that sense of final, irremediable loneliness which she constantly felt "here"—she pointed not at her heart but at her throat.

"You might stop talking German till we get started," said one of the other prisoners in our compartment, "or they'll take us for fascists." This was a Party worker from Vologda, whose name I don't remember. Her voice was hoarse and she spoke with a strong local accent. Her long, thin face with its cracked lips looked like a charred piece of firewood. Only the fine, light northern hair curling around her temples distantly recalled the fact that she had once been a woman. She was in a frantic state of mind and talked incessantly, quoting some figures about a milk delivery plan and denouncing the misdeeds of a certain Voskoboynikov, who had "put the quota too high." She spoke as though we were familiar with all the circumstances and the people involved in her case, breaking off from time to time with a cry of agony. She had been kept standing for long days and nights and still suffered terrible pains in her legs.

The fourth person in our compartment was Julia Karepova, the one with whom I had traveled from Kazan to Moscow. She did her best to out-talk the woman from Vologda and bring the conversation around to Kazan.

But now our train was shunted from the siding onto the main lines, and through the barred windows we could see a little life—that fascinating ordinary human life we had not seen for so long. What we saw was a corner of the Northern station. A train pulled in, and out of it poured a motley crowd of returning holiday-makers with their children, carrying bunches of flowers and smiling. Their belongings were not the miserable little bundles we called our "things," but the sort of pleasant, touching objects we had left behind—baskets of fruit, small suitcases, toys . . .

We watched them from the window, spellbound. The guard for some reason did not interfere. The people on the platform noticed us. A girl in a bright dress clutched her companion's arm, looking at us wide-eyed. We caught the words "Trotskyists . . . real live Trotskyists." She must have been telling him it was the first time in her life that she had seen real Trotskyists.

Then a woman with two small boys walked past, and I felt ready to die of envy and amazement. So there were still people in the world who were allowed to hold their children's hands.

At last our train started. Greedily I drank in the sight of the Moscow suburbs as they drifted by. Every station had its red banner printed with a slogan. All the slogans spoke of sabotage: "Liquidating the effects of sabotage in the transport system, we pledge ourselves to . . ." Here was a store with the slogan: "We shall stamp out sabotage in the retail-trade network," and on a power station: "Liquidating the effects of sabotage in industry, we resolve to exceed the norm. . . ."

"Julia! Have a look! There's been sabotage in every single branch of the economy!"

"Not so loud. . . . There's the guard. . . ."

The train was now speeding through the open countryside. It was the latter half of August. We caught the smell of the fields. Birds perched on telegraph cables like notes on a lined sheet of music. We were going to Yaroslavl, a light-blue town, cool and clean. I had been there in 1934 with my husband. Now it would be the color of lead. The wheels rattled—"with e-ve-ry step, with e-ve-ry step."

Solitary confinement for ten years. Day after day, August after August. My sons would be almost grown-up, and I—an old woman. Day in, day out, I would hear no human voice except for five commands a day: reveille, hot water, washroom, exercise, lights out. I would forget how to talk. I would forget the color of the sky and of the Volga. Cells for solitary confinement all had rats.

I had visions of Monte Cristo, Princess Tarakanova, and the infant Tsar Ivan VI.* The train whistled merrily. Carola groaned and muttered in German in her sleep.

We arrived at the golden hour of sunset. Again we stopped at some out-of-way siding. There was no platform and we jumped straight out onto the damp, dark-yellow sand smelling sweetly of childhood.

For some reason, the Black Maria had not arrived. No one had come to meet us. The guards fidgeted and whispered. Smiling happily, we sat on our bundles, greedily

---

* Princess Tarakanova was an 18th-century pretender, imprisoned in the Peter and Paul Fortress. Ivan VI Antonovich, infant Tsar 1740-41, was subsequently imprisoned and killed.

drinking in the fresh Volga air. So the prison system didn't always work without a hitch! For ten whole minutes we sat waiting for transport, staring at the open sky and spellbound at the sight of a gull from the Volga.

There seemed to be no end to the unexpected. Suddenly, it turned out that we were not going by a Black Maria but in an ordinary open truck. This was almost impossible to believe. Creatures of the underworld, living our unnatural lives in prison cells, were we to be allowed a glimpse of city streets and free people walking about them?

Julia was quick to voice her optimism. If they were fetching us in an open truck, the prison couldn't be as bad as we had thought, and the Butyrki rumors about the recent tightening of discipline must be nonsense.

"Everybody on board!"

We were driving through the streets of Yaroslavl. I recognized the hotel where Paul and I had stayed four years before. There were a lot of people strolling along the embankments. We saw the Volga, and took deep breaths of the pure air which had to last us a long time.

Carola's beauty and unusual dress drew the eyes of the passers-by. They looked at us with curiosity and a few smiled.

"Hello, girls!" shouted a tall young man, walking past with a group of friends. They waved their caps and I felt a surge of warm affection for these unknown people. How good that they were left in peace, and every evening strolled on the embankment.

Our truck swerved sharply to the right into the yard of a large prison. So this was the notorious Yaroslavl prison, Korovniki.* But we were no ordinary prisoners. We were

---

* Korovniki: the word in Russian means cow sheds.

state criminals of special importance. So we were escorted to the isolation block separated by high walls and watch-towers from the rest of the prison.

We crossed the threshold, to be buried alive for a little more than two years.

33

• *Five steps by three*

To this day, if I shut my eyes, I can see every bump and scratch on those walls, painted halfway up in the favorite prison colors, brownish-red and a dirty white above.

Sometimes in the soles of my feet I still feel this or that crack in the stone floor of my cell: Number 3 on the second floor, north side.

And I still remember the physical anguish, the despair of my muscles, as I paced the area in which I was henceforth to live. It was five paces long and three across. At a pinch, by taking very short steps, I could stretch it to five and a quarter. One, two, three, four, five. Round about turn on my toes—not to waste any space—and again, one, two, three, four, five.

An iron door with a flap-window and a peephole. An iron bunk screwed to the wall and, against the opposite wall, an iron table and a folding seat—very uncomfortable to sit on but well placed for observation through the peep-hole. Nothing but stone and iron!

The window, facing north, and high up as in all solitary cells, had been fitted with sturdy bars by the late Nicholas II, frightened by the revolution of 1905. But someone even more frightened than Nicholas had provided it, over and above the bars, with a high, solid wooden screen which ensured perpetual twilight.

Above the screen, what remained of the pale-blue Yaroslavl sky looked like a narrow trickle of water. Even this was often blotted out by the crows. For some reason, a flock of these ill-omened birds circled around the prison as if in search of prey. Winter or summer, we were never rid of them. And when I picture to myself the window of my cell in Yaroslavl, I invariably see it with its black border of crows perched on top of the screen.

I was taken out of my cell three times every twenty-four hours: morning and evening to the washroom, and before or after dinner for exercise. What a blessing that my cell was so far from the washroom! It meant that I had to walk almost the length of the corridor. This was in the form of a gallery which ran around the stair well. The well was protected by thick wire netting so that no one might take it into his head to jump from the second floor and die on his own initiative instead of at the moment laid down by higher policy.

The corridor was covered by a magnificent plush carpet into which one's feet sank noiselessly. Going to the washroom, I walked as slowly as possible, making a show of weakness quite plausible in a prisoner held in solitary confinement. I tried to use every second of the way to take in all I could of my surroundings with my trained, observant prisoner's eye. For, compared with the cell, the corridor was a whole enormous world of its own.

I doubt whether Sherlock Holmes himself could have

observed that tiny corner of the world more closely. I became an expert in his "inductive method," and every visit
to the washroom added to my mental picture of the prison.

A large wooden bin into which uneaten crusts of bread
were thrown stood by the window of the corridor. This
was very different from Butyrki! There, the very notion
of "uneaten bread" would have been laughable. But here
there was bread to spare. Prisoners in solitary had no appetite, and I carefully noted the daily quantity of bread
which was thrown away. Sometimes a whole portion
seemed to have remained untouched. Was someone perhaps on hunger strike?

Another time, I would notice that the door of the cell
opposite mine was open, apparently while the occupant
was being given his exercise. I saw with envy that it was
a better cell than mine: it faced south and got a little sunshine in spite of the wooden screen. In mine, the bottoms
of the walls were decorated with a thick pattern of mold.
A sure guarantee of rheumatism.

The walk was the focal point of the day and was attended with as much regal ceremony as though Mary
Queen of Scots, at the very least, were taking an outing.

About a quarter of an hour before the appointed time,
the window in the door flapped open and the warder
poked his head in.

"Get ready for your walk," he whispered, softly and
mysteriously as though someone next door were dying.

I dressed, breathlessly awaiting the longed-for sound of
the key turning in the lock. Then, the warder in charge
of my end of the corridor handed me over to another who
led me to the head of the stairs. There, a first-floor warder
met me and escorted his "dangerous political prisoner" to
the ground floor. Finally, the ground-floor warder took

me out into the small exercise yard, overlooked by a watchtower, where yet another warder stood and watched me, without taking his eyes off me for one moment during the whole of my walk.

Thus, five strapping young peasants, clearly intended by nature to fulfill production plans on farms and in factories, were employed in exercising one dangerous "terrorist"! Their faces showed nothing but proud consciousness of the importance of their duties and of the trust reposed in them. I could well imagine what their political instructors had told them about us!

The exercise yards for prisoners in solitary confinement were in effect nothing more than roofless cells. The asphalted courtyard had been divided into half a dozen pens, each about fifteen yards long. The walls and the asphalt underfoot were a dirty gray. There was not a blade of grass. Though always quite alone, I had to walk with my hands behind my back. After pacing up and down for ten or fifteen minutes, I was returned to my cell, the five warders handing me on, as in a relay race, in reverse order.

Yet I recall my walks, such as they were, almost with fondness. They were brief intervals of life. I waited for them impatiently. I remembered them in the evening. To be deprived of exercise—and this was a fairly frequent punishment—was a horrible misfortune. After all, fifteen paces were more than five. And there was the sky. . . .

To my dying day I shall not forget the clear, high-vaulted Yaroslavl sky. No other town has anything to compare with it. And besides, one could sometimes see gulls flying over from the Volga. . . .

And the ships' sirens! How can words convey what they meant to a prisoner in solitary confinement, especially to one who, like me, had lived by the Volga? I heard them as the living voices of friends. For I knew those ships, every one of them—the proud white swans of the old Samolyot Company, and the busy, fussy tugs trailing their barges, and the shrill-voiced excursion steamers. . . .

It could never have occurred to my warder how many new impressions, dreams, sweet memories, a prisoner brought back from a fifteen-minute walk in a gray roofless cell.

Exercise made me hungry, and I managed, somehow, to eat my dinner. The quantity was enough to keep one actually from starving to death, but the quality was such that one could scarcely manage to live on it, either. There were no vitamins in the food. For breakfast, we were given bread and hot water with two lumps of sugar; for dinner, soup and *kasha* without fat of any sort; for supper, a kind of broth smelling repulsively of fish oil. *Kasha* was of various sorts—wheat, barley, millet, but, more often, the large-grained barley known in Butyrki as "shrapnel." Soup was always buckwheat.

According to the rules displayed on the wall, books were allowed at the rate of two every ten days. But throughout the first month of my stay, the library was closed for stock-taking. So I had sixteen hours of leisure a day to fill in as best I could. I tried to establish some sort of routine to stop me from going mad. The most important thing of all was not to forget how to talk. The warders were trained to silence; they spoke only half a dozen key words a day—reveille, washroom, hot water, exercise, bread. . . .

I tried to do gymnastics before breakfast. The flap-window clicked open:

"That's forbidden."

I tried lying down after dinner. Another click:
"You mustn't lie down except after lights out, from 11
P.M. to 6 A.M."

So what was left? Poetry . . . only poetry . . . my
own and other people's.

And so I paced my five steps up and three steps down,
composing lines like these:

> Between these walls of stone
> All roads are just as short:
> By any count this cell
> Is never more than three by five.*

No good without a pencil! I would never have made
an *akyn*.

After dinner was my time for Pushkin. I gave myself a
lecture about him, then recited all I could remember of his
poems. Like a chrysalis transformed into a butterfly, my
memory, cut off from all outside impressions, suddenly
blossomed. Wonderful! I found I even knew the "Cottage at Kolomna" by heart. That would do nicely till
suppertime.

But the time after supper was the worst. The silence
thickened, became tangible and stifling. Depression attacked not only the mind and heart but the whole body.
Even my hair seemed to bristle with despair. I would have
given anything to have heard just one sound.

The worst thing of all was that, ever since the torture
by insomnia on the "conveyor belt," I was incapable of
deliberately making myself go to sleep. The knowledge
that the hours allotted to sleep were running out and could
not be made up for by day drove me to desperation. Ter-

---

* This and subsequent renderings of Mrs. Ginzburg's verses are free and
sometimes abbreviated renderings which do not pretend to reproduce
the fine quality of the original.

rified of wasting the precious time, I frantically tried to sleep, and this drove sleep away altogether. There was nothing for it but to recite or make up verses of my own, with such meager results as the following:

Silence
Every rustle
Whisper
Step
Burns
And deafens
Like a shot.
Silence unravels
Like yarn.
Is it my heart?
Or the guard?
Or a mouse?
The soul
Is taut as a membrane.
Every footstep
Hammers and bruises.
Night is close
And cannot be put off,
Close and stifling
Like cotton wool.
All my losses
Are here beside me.
What to trust
When all is lies?
Every rustle
Cuts like a knife.
Now it stops.
Who is left?
What awaits me?
The night grows wider
And dreams more bitter.
How much silence
Is there in the world?

34

• *Major Weinstock's twenty-two commandments*

They hung on the wall, just above my bunk. As there were still no books, the twenty-two commandments were the only printed words available to me. I studied them to the point of stupor.

The opus was divided into three unequal parts: "Prisoners are obliged," "Prisoners are allowed," and—by far the longest—"Prisoners are forbidden."

Prisoners were obliged to obey all rules and orders without question or delay. They must clean their cells on the appointed days and carry out their slops twice daily.

Prisoners were allowed (in principle, though in practice the permission of the prison governor had to be obtained every time) to exchange two letters a month with their closest relatives; this meant no one but parents, husbands or wives, and children. From those same relatives they could also receive up to fifty rubles a month, to be spent at the prison store. Prisoners were allowed to take exercise in the prison yard, the period to be fixed by the prison governor, and to borrow two books from the library every ten days.

This was the limit of Major Weinstock's generosity. The list of prohibitions, on the other hand, had been compiled with great thoroughness and attention to detail. It was forbidden to go near the window or to sit with one's back to the door; to make marks in the library books or to tap messages on the walls. It was forbidden to talk (with

whom?) or even to sing in the cell. And so on, and so
forth. . . .

Finally, the penalties for infringing these rules were
listed in loving detail: withdrawal of the privileges of exercise, correspondence, use of library and prison store; confinement in a punishment cell, and, if all else failed, a
court trial.

The document was signed "Head of Prison Administration (State Security Organization), Major Weinstock,"
and countersigned in the top left-hand corner: "Approved:
Commissar-General for State Security, Yezhov."

Yet our one desire was that these rules should not be
changed—I realized this two years later during my transfer to Kolyma. All we longed for was that things should
"get no worse," for every day brought fresh signs of the
growth of terror and lawlessness. Someone's devilish ingenuity was ceaselessly at work to find chinks in the walls
of our dungeons and ways of stopping them.

Every day brought some new refinement. Only yesterday the window at the end of the corridor was merely
whitewashed; today it had a dark shutter. Only yesterday
the warder took no notice of my sitting with my back to
the peephole; today he opened the flap-window and hissed
angrily:

"Turn around!"

My walks became shorter and shorter, permission to buy
food from the store was given less and less often. Worst
of all, the prison governor was replaced.

The old one was still there when I arrived, and he came
to see me the day after my arrival. I had heard about him
in Butyrki. He was typical of the political-prison administration prior to 1937, before the prisoner's destruction became the object of the operation.

Good-natured and round-faced, he half opened the door, said "May I?," came in, introduced himself, and asked if I had any complaints or requests. He assured me that the library would soon reopen, took down my request to correspond with my mother, and went off, leaving a general impression of decency. His every word and gesture implied: "I work here but I don't enjoy it. If there's anything I can do for you, I gladly will."

Alas, it was to be our first and last meeting. The decision of the July plenum to tighten the prison system was about to be put into force.

Five or six days later the door flew open and a swarthy man in uniform came in. In walking he bent at the knees like a camel. He looked right past me as he spoke.

"I'm the new governor," he snapped. "Any questions?"

His features and expression were those of the human vulture in a Georgian film, the one who does to death the dovelike heroine, Nata Vachnadze. I immediately christened him Vulturidze and the name stuck. We used it when we talked about him later, on our journeys, and many prisoners thought it was his real name.

He forced his words out through his long, clenched teeth, with an expression of deep disgust.

"Any ques-stions?"

"Yes. Tell me, how long am I to be kept in solitary confinement?"

"Don't you know your own sentence? Ten years."

After this one exchange, I always said I had no questions. What was there to ask? Everything was clear.

But life administered its own correctives to Major Weinstock's rules and to Vulturidze's promise of solitary confinement for ten years. More and more batches of prisoners arrived from Moscow. The prison was bursting at the seams. Contrary to the spirit of the twenty-two commandments, extra bunks were placed in some solitary cells.

One day the tomblike silence was broken by the clanking of keys and by muttered conversation in the corridor. I guessed what this meant, and in painful suspense awaited my girl Friday. When she arrived, it proved to be one of those miracles of which one says: "It couldn't happen in real life." But it did. Of all people, my cellmate turned out to be Julia Karepova, the one with whom I had traveled from Kazan to Moscow and from Moscow to Yaroslavl.

35

• *Bright nights and dark days*

For twenty hours a day we talked till we made ourselves hoarse. The fact that we were human beings, capable of articulate speech, filled us with excitement and pride.

Before long I had heard in detail not only Julia's life story but that of all her relatives for the past three generations. I used to recite poetry to her for six hours a day, and we rehashed even the stalest Butyrki gossip.

Afterward, a reaction set in. We suddenly ceased to talk and became absorbed in our own thoughts, wondering what was in store for us. The more one brooded about it, the more likely it began to seem that we should not leave prison alive.

We were saved unexpectedly from these gloomy thoughts. One day, the flap-window opened suddenly and a sort of folder was thrust through it, looking rather like a class register. It was held by a tow-headed warder we knew as Yaroslavsky, and in whom for once kindness seemed to prevail over routine severity. With a broad smile he pronounced the magic word "Catalogue!"

It was an object lesson in not losing hope. We had long since come to the conclusion that the library "stock-taking" would last out our ten years, but lo and behold—this really was the catalogue, and a good one at that. The library was a large one, with an excellent choice of books.

This was the end of our loneliness. Tomorrow at this hour I would have visitors: Tolstoy and Blok, Stendhal and Balzac. How stupid I had been to have thoughts of death!

In a flurry of excitement, making mistakes in our haste, we wrote down the numbers of the books we wanted. Next day we would each get two from our list. What luck that I was no longer alone, so that the books would arrive four at a time! This was a ration one could live on. . . . Yaroslavsky himself was so affected by our radiant happiness that, after looking furtively around, he bared his irregular but remarkably white teeth in a broad grin and, nodding approvingly, said "Tomorrow!"

Next day, then, I held in my hand four books and felt almost faint with greed, unable to decide which of them it would be least painful to leave to Julia, who had good-naturedly allowed me to choose first. Which would I start on? Of course, Tolstoy's *Resurrection*! After some thought, I let Julia have a selection of Nekrasov's poems, and she was soon exclaiming with surprise:

"I always thought that the wives of the Decembrists *
endured the most frightful sufferings, but listen to this:

Of wondrous build, so light, so strong,
The little wagon speeds along.

They ought to have tried one of our Stolypin coaches!
. . ."

But there was no time to talk now—only to read, and I
was soon buried in the much-thumbed volume of Tolstoy.

At home, I had always been looked on as a passionate
and indefatigable bookworm; but it was here, in my stone
sepulcher, that I really explored for the first time the in-
most meaning of what I read. Up till then I had skimmed
the surface, enlarging my mind in breadth but not in
depth. And after I came out of prison I found I could no
longer read as I had done in my cell at Yaroslavl, where
I rediscovered Dostoyevsky, Tyutchev,† Pasternak, and
many others. There, too, I worked through several books
on the history of philosophy, a subject I had never studied
before. Strange to say, one could obtain freely in prison
a number of books which had long since been withdrawn
from public libraries.

It is a commonplace to explain the profound effect
which books have on a prisoner in solitary confinement by
the absence of impressions from outside. But this is not
quite the point. When a human being is isolated from the
"rat-race" of everyday life, he achieves a kind of spiritual
serenity. Sitting in a cell, one no longer has any call to

---

* Wives of leaders of the abortive uprising of 1825 who voluntarily followed their husbands into Siberian exile. Nekrasov wrote a poem
about them.
† Fyodor Tyutchev, 1803–1873, Russian poet.

pursue the phantom of worldly success, to play the diplomat or the hypocrite, to compromise with one's conscience. One can immerse oneself in the lofty problems of existence, and do so with a mind purified by suffering.

If even the labor camps, with their savage animal struggle for survival, allowed thousands of our fellow citizens to keep their integrity, how much more is this true of those in solitary confinement! Its ennobling influence is unquestionable, provided it does not go on so long as to undermine the foundations of personality.

How often later on, when I was in the camp, I used to remember with affection the dreadful days of solitary confinement at Yaroslavl! For, despite the horrors of life there, no place at any time before or since has so developed the better sides of my personality. During those two years I was far kinder, more intelligent and perceptive than at any period in the rest of my life.

Even the daily increasing severity of the prison regime could not damp the joyful excitement which we felt at the opening of the library. If only they would not shut it again! Meanwhile we bore stoically even the process of putting us all into prison clothes, the so-called "Yezhov uniform." Our own clothes were taken away and we were issued with grayish-brown sateen skirts, blouses with a diagonal crisscross of brown stripes, and jackets of stiff material with a similar "convict's" pattern. Caps and boots were in short supply, so I was able to keep my flowered scarf and down-at-heel red slippers, which were now the only patches of color in our life. The block warder, whom we called the "Nabob," remarked mockingly, as he stuffed our best overcoats into a sack, that this was the end of our fine feathers.

At first we took this new development tragically. No

woman of thirty likes to be transformed into a scarecrow, even if no one is going to see her. But we soon turned our minds to the problem of how to keep at least one brassière each. Prison underwear consisted only of coarse calico shirts and trousers. There were no brassières, and the resultant slovenliness of our appearance was more than we could bear. So each of us managed by some sleight of hand to hide away a brassière, which we held on to despite the searches which took place in this prison twice a month. We washed them over the slop bucket and darned them with the help of fishbones which we extracted from the evening stew. I should have kept mine as a memento of the indestructible Eternal Feminine, but it got lost during one of our numerous moves from one camp to another.

Within a week of the first issue of books, we were both suffering from eyestrain. Our cell got virtually no light in the daytime, what with its looking north, the solid wooden screen, and the huge black crows that perched on it. If we went on reading for eight or ten hours a day, it was clear that we might go blind. We had to do something about this situation.

Although the authorities were careful to shift the warders about from one corridor to another, so that we should not get used to them or establish human relationships, the same ones came back to us from time to time and we learned to distinguish between them. Each one was sized up, and each was given a nickname.

On days when the "Nabob" was patroling our corridor, or when Worm—a nasty, thin-lipped, pimply type—was prowling around the peephole, we observed instructions religiously. But when Yaroslavsky, or St. George, or the Dumpling—a round, pleasant-faced woman—was on duty, we turned day into night and vice versa.

Having learned to sleep in a sitting position, we sat at an angle to the peephole in attitudes that suggested we were reading. Our books lay open before us, but in fact we were sleeping soundly. At night, however, when the cell was filled with dazzling electric light, we contrived to hold the book under the blanket and thus read till nearly dawn. Of course our eyes still suffered, from the awkward posture and from lack of sleep, but it was better than before, and for a long time we managed to fool the guards. Occasionally, however, the flap-window opened and we were admonished in these terms: "Number 1, wake up Number 2 and tell her not to hide her face." This meant that Number 2, which was Julia, had pulled the blanket too far over her head.

Such were our black days of sleep and our nights of blinding light and surreptitious reading. Such was our life of physical and mental suffering, of transfigured hours in the society of books, of alternate hope and despair.

There came a time when the sky, as we looked up at it from our exercise pens, grew grayer; fewer gulls flew across it, but the crows perched ever more thickly on the window screen. Autumn was upon us.

36

• *Captain Glan's dog*

In one of Knut Hamsun's novels there is a character called
Captain Glan with a dog answering to the name of Aesop.
Although the whole spirit of our prison life was permeated by "Aesopian language," * Julia, clearly overesti-
mating our warders' education, was frightened of using
this expression aloud. Thus it was that we came to use the
term "Glan's dog" for the methods of allegory, fable, and
double-talk in which we became adept for the purpose of
conversing on forbidden topics and, especially, corresponding with the outside world. This meant with my mother,
since Julia had no close relatives and was therefore de-
barred from writing.

The writing of my fortnightly letters was a momentous
event for which we prepared long beforehand, weighing
every word.

It was a difficult task to write something that would be
fully intelligible to my mother without attracting the vigilance of the prison censor, who would return without
compunction any letter in which he detected the slightest
ambiguity. This happened, for instance, when I asked my
mother, in all sincerity, to make sure that Vasya was
vaccinated. Any reference to illness was suspected of containing a hidden meaning.

We wrote with regulation pencils made of Bakelite with
lead refills that needed no sharpening: we were not allowed

* Traditional term in Russia for "coded" language.

any sharp objects such as penknives. The letters were put in envelopes in the censor's office.

In order to give Mother as much news as possible and to find out about my husband and children and all our friends and relatives, I used a device we had invented of writing in the third person. This involved long preparation, including the choice of a code name for myself. We eventually decided that I should be called "Eva" which sounded a little like "Genia"; and in due course I wrote a letter containing the cryptic sentence:

"Don't worry so much about the children, especially little Eva. After all, she's not alone, and I'm sure her aunt looks after her quite well."

My mother caught on at once and replied that indeed she hoped that Eva would be all right, but didn't I think her aunt was a bit strait-laced? Would she allow the child to get about and see her friends? This was in order to find out whether I was in solitary confinement. . . . From this point on, our system worked beautifully. By writing about ourselves as though we were children we got all kinds of strictly forbidden information past the censor without arousing his suspicions. Thus, when Mother wrote, "Paul hasn't taken his exams yet," I understood that my husband had not yet been tried and sentenced. The fact that my sister's husband, Shura Korolyov, had been disgraced and arrested was conveyed as follows. First my mother wrote: "Shura has a new job, he's working in a garage." Since he had been a history professor, this could only mean that he had been expelled from the Party. In her next letter she wrote, "Shura's staying with little Paul now," which could only mean one thing.

We corresponded in this way for two years. From Mother's full accounts of my children's doings I concluded that they were all right, and this gave me strength to endure everything. Long afterward, at Kolyma, I learned that at the time when Mother had written, "We had a New Year's party for Vasya," he had in fact been in a home for prisoners' children, where he was registered under the wrong surname. For months the family despaired of finding him, until my husband's brother succeeded, in 1938, in tracing him at Kostroma. It was a mercy that I knew none of this at Yaroslavl.

We also used "Aesopian language" in our notes and conversations. We were allowed to buy two notebooks a month from the prison store and to write in them anything we liked. But since they had to be handed to the censor as soon as they were full, we could not use them, for example, to write verses in as we should have wished.

Our prison experience illustrated the fact that any human being in the position of Robinson Crusoe will, as it were, retrace the development of the species, passing through the various stages of technical progress. We made a needle out of fishbone, and thread out of our own hair. We invented a system of shorthand for writing verses (I have completely forgotten it now) and brought to exquisite perfection the technique of wall tapping, which in the awful stillness of this place was far more dangerous than in the cellars at Kazan.

When I had written my verses I used to learn them by heart, then rub them out with bread crumbs and substitute algebra sums or conjugations of French verbs. But the main object of our subterfuges was, of course, to overcome the strict isolation from the world and from one another which was one of the prison's basic laws.

From the authorities' point of view, the ideal was that each of us should feel as though she were the only inmate of the prison. Since they had had to put two of us in one cell, I was at least assured of the existence of Julia Karepova. But who were our neighbors? By listening intently to the vague sounds that came through the walls on either side of us, we decided that they were both "solitaries," without an extra bunk in their cells. No doubt the more important prisoners were kept as long as possible without company. . . . Our neighbor to the right paced incessantly up and down, the creaking of her heavy prison boots being audible even through the three-foot wall. When we asked her name and how long she was in for, she countered by asking which party we belonged to. When we replied that we were Communists, she retorted, "No member of that Party is a friend of mine," and banged the wall with her fist, after which she ignored us for fully two years. She was evidently a Menshevik or a Social Revolutionary of the same stamp as Mukhina at Kazan.

With our neighbor to the left, on the other hand, we were able to strike up regular communication. Almost every day we exchanged "telegrams" composed in such a style that the authorities would not have understood them even if they had intercepted them. Her name was Olga Orlovskaya; she was a journalist from Kuibyshev and the ex-wife of a certain Lenzner, who had played an important role in the Trotskyist opposition. Although she was a keen Party member and had been divorced from Lenzner for many years, she was arrested for association with him. She had now been several months in a solitary cell and was delighted to be in contact with us. We "spoke" to each other during mealtimes, when the silence was broken by the clatter of ladles and tin bowls. Our chief topics of discussion were furnished by the local newspaper,

the *Northern Worker,* to which we were allowed to sub-
scribe out of the fifty rubles which our families had
sent us.

What a newspaper it was! If any reader were to pick
it up today, he would think he had gone mad. The process
of rooting out "enemies of the people" had become quite
stereotyped. For instance, one would read an article about
the slackness of a district committee secretary who had
stated that there were no culprits left to arrest in his district. The writer expressed indignation at such softness
toward "hostile elements" and cast doubt on the secretary's own loyalty. Several times a month, full-page spreads
would be devoted to the trials of local leaders. The drab
columns bristled with bold-type references to the "supreme
penalty" and to "sentences carried out," interspersed with
fulsome praise of "true sons of the people" and "simple
Soviet folk."

The elections to the Supreme Soviet were about to be
held, the first under the new constitution. The candidate
for Yaroslavl was Zimin, the first secretary of the regional
committee, whose predecessor in that office had just been
arrested. Every issue of the newspaper contained photographs of Zimin in various poses and eulogistic accounts
of his services. A few months after the elections, Zimin
was arrested together with the entire bureau of the regional committee, and the *Northern Worker* devoted pages
to denouncing "the arch-spy Zimin, who wormed his way
into the higher command of the Party."

We got the impression that whole layers of society were
being eliminated even before Kaganovich * used this very
term in noting with satisfaction that the government had

---

* Lazar Kaganovich, b. 1893. Bolshevik leader, close associate of Stalin.

"liquidated several layers" in its campaign to stamp out the after-effects of sabotage.

Olga and I discussed all these matters through the wall in "Aesopian language." Her replies testified to her keen intelligence and journalist's flair for summing things up in a vivid phrase. We communicated in this way for two years. But only in 1939, on the way to Kolyma, did I discover that in spite of everything she worshiped Stalin. In her solitary cell at Yaroslavl she had written a poem to him which began:

Stalin, my golden sun,
If death should be my fate,
I will die, a petal on the road
Of our great country.

This was not really out of the ordinary, for I met many people in the camp who managed to combine a shrewd sense of what was going on in the country at large with a religious cult of Stalin.

Julia, who had a taste for detective stories, got so carried away by her "Aesopian" techniques that even I could not always understand her intricate allusions. She became even more careful after Olga warned us that she had been deprived of books as a punishment for certain underlinings which she had not in fact made. From then on we scrutinized all our books with the utmost care before returning them.

Julia's language became still more "Aesopian" after she got it into her head that a microphone might be concealed in a niche in the wall above her bed. I tried vainly to convince her that this would hardly be necessary from our jailors' point of view, since we had already been investigated, tried, and sentenced as enemies of the people, and they could not expect to get any further evidence out of us. Julia continued to be terrified of "Nicky" (her name for the niche), and invented such grotesque ways of preserving the secrecy of our conversations that I sometimes burst out laughing, hiding my head under the straw pillow so as not to attract attention.

But in spite of all our efforts and precautions, the heavy hand of the prison authorities fell on us at last. The punishments which they dispensed, like our sentences themselves, were not proportionate to any misdeeds of ours but were meted out according to a plan, a prearranged formula. And in this plan a key date was the first of December—the third anniversary of Kirov's murder.

37

• *The underground punishment cell*

As usual, misfortune struck when we were least expecting it: indeed, we were feeling particularly cheerful that day.

In the morning they had brought us food from the prison store: a pound of sugar, a quarter of a pound of meat, and some cucumbers that had found their way there. These were yellow, wizened, and frightfully bitter. Julia, with her background of old-fashioned housekeeping, was carried away by an ingenious plan for pickling them.

"It's perfectly sensible! We can ask the day and the night warder, and the one tomorrow morning, for a pinch of salt each. For the next three days we won't take any hot water, or if we do, we'll pour it into the slop pail. Then we can salt them in the hot-water jug, and in three days' time they'll be delicious."

"Honestly, Julia, I ought to write a poem in your honor."

"Yes, I think you might. You write about everything under the sun except your faithful cellmate."

So I tried my hand at classical elegiacs:

Not in iambics I sing thee, nor yet with the frivolous trochee;
No, elegiacs of old fitly echo thy praise.
E'en though Vesuvius again should erupt and astonish the nations,
Thou on its summit shouldst yet pickle and salt with a will.

Just then, as we were bubbling with suppressed laughter, the key turned in the door. My heart jumped into my throat: any visit at an unscheduled time could only mean trouble. It was the block warder, ordering me to go with him. There had been a few such occasions since our arrival, when we were taken to be fingerprinted or to the dentist. But for the dentist one had to apply beforehand, so what could this be?

As I walked down the stairs I could hear Julia coughing to express alarm and sympathy. We passed the first floor and still went down and down. Clearly this was more than a mere formality. The continuing descent was ominous. How many underground floors could there be?

Finally we came to a halt in a sort of narrow dungeon, where I saw the stout, stocky figure of the senior warder we called the "Nabob." He had a dark, swarthy face with startlingly pale eyes, and spoke with a strong Ukrainian accent. Having checked who I was, he pulled out a register and announced that the governor had given orders for me to be confined in an underground punishment cell for five days "for continuing counter-revolutionary activity in prison by writing her name on the washroom wall."

The charge was a manifest lie: I had not written my name, and should have been very foolish to do so. We knew perfectly well that each time we left the washroom the warder who mounted guard outside looked in to make sure we had not planted a bomb there. His other duty was to hand each of us a single square of newspaper, which responsible task he performed with a suitably solemn and impassive air.

All this flashed through my head, and I tried to point out to the Nabob that no one but an idiot would try to write on the wall in these circumstances. He paid no attention, but ordered me to confirm by signing the book that I had been informed of the governor's instructions.

"Certainly not! I won't sign any such lying nonsense. Anyway, what's all this about counter-revolutionary activity?" As I quoted these words from the charge, I suddenly realized their point. I saw in my mind's eye the passage in Major Weinstock's twenty-two commandments which said that prisoners who continued counter-revolutionary activity were liable to a new trial. If I had acknowledged the order, it would look as though I were confirming its truth, and I should be giving them an excuse to re-try me and probably, this time, condemn me to death.

"I won't sign it. It's a trick!"

"All right, don't, it'll be the worse for you. Come on now, get undressed."

"What!"

"Undress, I said. You have to wear special clothes in the punishment cell. Come on, look sharp!"

He urged me forward into a sort of triangular stone space. There was no lamp or window, and the only light came from the open door. There was a damp and chilly draft: clearly the place was not heated. A narrow plank bed was fixed a few inches above the floor, and on it were the rags I was supposed to change into: the greasy remnant of a soldier's overcoat, and enormous bast sandals.

"I won't!"

"Oh, yes, you will, or we'll put you into a worse place," said the Nabob in sudden rage, and before I knew what was happening he started to undress me by force. I felt his paws on my breast. I heard myself scream wildly, and broke loose from him. This time it was more than I could stand. I shouted and struggled even more desperately than I had in the Black Maria after my trial. Then I had hit my head against the wall, seeking to harm no one except myself; but this time I was so beside myself that I tried to fight the Nabob, who could have felled me with one blow of his fist. I scratched and bit and kicked him in the stomach, uttering frightful words:

"You fascist scum—just you wait, you'll be punished one day!"

Suddenly I felt a sharp pain, so excruciating that I almost lost consciousness. The Nabob had twisted my arms back and bound my hands together with a towel. Through my daze I saw a wardress rush over to help him. She undressed me down to my prison shirt, even taking out my hairpins. Then everything became a blur, and I sank into a black abyss where everything seemed on fire.

When I came to myself I was freezing with cold. The

toes of my left foot were so numb that I could not feel them. I had in fact got second-degree frostbite, and to this day my foot swells and aches every winter.

My whole body throbbed with pain. I lay on my back on the low plank bed, almost naked, with my shirt and the grubby coat over me. But my hands were unbound—the wardress must have had the decency to do that before she threw me in here.

I looked into the darkness but could see absolutely nothing, not a glimmer of any kind. Then I heard the tramp of military boots approaching. The little window in the door was unlocked, and a blessed stream of light flowed in. At least I could see, I was not blind! Now that I knew this I should be less afraid of staring into the dark.

"Water," said my guard, offering me a dirty, rusted mug. The water was coated with a film of grease. I swallowed two mouthfuls and used the rest to wash with, wiping my face and hands meticulously and drying them with the top of my shirt. Now I was once more a human being and not a grimy, hunted animal.

"Here's some bread."

"I won't eat it."

"Why?"

"It's too filthy to eat here."

"I'll tell the supervisor."

He went away, but did not shut the window quite so tightly as before, so that a gleam of electric light was visible around the edge. I fixed my eyes on it and felt immensely relieved.

What I must do was to keep count of the days and nights so that they did not all merge into one. They had just offered me bread: this would be the first day, and so

I made a rent in the hem of my shirt. When there were five of them I would be let out.

How palatial our cell would seem! I wondered if they had treated Julia like this too. She had trouble with her lungs as it was. . . .

It was impossible to sleep because of the cold and the rats. They scuttled past my face and I clouted them with the huge sandals. What was I to do? Of course—there was always poetry!

I recited Pushkin, Blok, Nekrasov, and Tyutchev. Then I composed a poem of my own—actually without a pencil! It was called "The Punishment Cell."

> This is no fantasy of a mad producer
> Or something dreamed by Edgar Allan Poe.
> As it dies away, I really hear
> The tread of my jailors' boots.
>
> In their drunken jackals' zeal
> How vicious they are, and vile . . .
> In my cold stone cell
> Darkness descends on me.
>
> Can a soul in hell be more lost?
> I must drink my cup to the dregs,
> But at least I am not alone
> In my calvary.
>
> A flagstone is my only cushion,
> But Pushkin, sitting in one corner,
> Sings me a song
> About Gurzuf at night.
>
> And, unseen by any guard,
> Another priceless friend
> Comes into my cell—
> His name is Alexander Blok.

Poetry, at least, they could not take away from me! They had taken my dress, my shoes and stockings, and my comb, they had left me half naked and freezing, but this it was not in their power to take away, it was and remained mine. And I should survive even this dungeon.

38

• *Comunista Italiana*

There were now four tears in the hem of my shirt: four times I had been offered bread, and four times refused it. I was beginning to get the hang of my new cell: I could recognize the Nabob's steps and his hoarse whisper, and the sounds when the guard was changed. I was able to deduce that the cellar contained no less than five "black holes" similar to mine.

In the same way, I at once recognized the loping, camel-like gait of the prison governor. When my door opened, I turned toward the wall so as not to look at his sneering face. He waited to see whether I would give any sign of life, and I waited for him to shout "Stand up!" But he addressed me with Olympian calm:

"Are you aware that hunger strikes are forbidden here?"

I said nothing. Even in my own cell I had refused to talk to this brute—and I certainly wasn't going to do so here!

"I repeat: do you know that in this prison, hunger-striking is regarded as a continuation of counter-revolutionary activity?"

I bit my lips hard and remained silent.

"You refused to acknowledge the detention order or to accept food. That is sufficient ground for me to make a report on your behavior to the judicial authorities. Do you realize this?"

He could go to the devil or shoot me on the spot, I was not going to utter a word. What had I to lose? Wouldn't I be better off in the grave than here?

He paused for another few moments, then turned on his heel (I remembered from his monthly visits how shiny his boots were) and disappeared. The door closed again, but suddenly the little window opened and I saw Yaroslavsky's friendly face with its rosy cheeks and sharp pig's bristles. He whispered rapidly:

"You'd better take the bread or they'll do you in, they really will. . . ." Then, though he had been about to say something more, he slammed the window to again. Someone was coming along the corridor.

I heard the sound of several feet, muffled cries, and a shuffling noise as though a body were being pulled along the stone floor. Then there was a shrill cry of despair; it continued for a long while on the same note, and stopped abruptly.

It was clear that someone was being dragged into a punishment cell and was offering resistance. . . . The cry rang out again and stopped suddenly, as though the victim had been gagged.

I prayed, as Pushkin once did, "Please God, may I not go mad! Rather grant me prison, poverty, or death." The first sign of approaching madness must surely be the urge to scream like that on a single continuous note. I must conquer it and preserve the balance of my mind by giving it something to do. So I began again to recite verses to myself. I composed more of my own and said them over and over so as not to forget them, and above all not to hear that awful cry.

But it continued—a penetrating, scarcely human cry which seemed to come from the victim's very entrails, to be viscous and tangible as it reverberated in the narrow space. Compared with it, the cries of a woman in labor were sweet music. They, after all, express hope as well as anguish, but here there was only a vast despair.

I felt such terror as I had not experienced since the beginning of my wanderings through this inferno. I felt that at any moment I should start screaming like my unknown neighbor, and from that it could only be a step to madness.

Suddenly the monotonous howl began to be punctuated by shouted words, though I could not make them out. I got up and, dragging my enormous sandals, crept to the door and put my ear close to it. I must find out what the wretched woman was saying.

"What's the matter? Fallen down or something?" came Yaroslavsky's voice from the corridor. He again half opened the window of my cell, and as the light gleamed I could hear words pronounced fairly clearly in a foreign language. Could it be Carola? No, it did not sound like German.

Yaroslavsky looked upset. It was indeed a beastly job for this peasant youth with his fresh face and porcine bristles. I am sure that if he had not been afraid of the accursed Nabob he would have helped both me and this woman who was screaming. At present the Nabob was evidently not there, for instead of slamming the window Yaroslavsky held it open and whispered:

"You'll be back in your cell tomorrow. Only one more night to go. What about a bit of bread, eh?"

I wanted to thank him for these words and above all for his kind expression, but I was afraid of scaring him by too much familiarity. So I only whispered: "What's the matter with her? It's terrible to hear."

He shrugged and said: "They haven't got the guts, these foreigners, they just can't take it. She's only just come in, and yet she makes all that fuss. The Russians are different, they don't kick up a row. Look at you for instance, you've got five days and you're not crying. . . ."

At that moment I heard clearly, in the midst of the wailing, the words "*Comunista Italiana, Comunista Italiana!*" So that was it! No doubt she had fled from Mussolini just as Klara, my cellmate at Butyrki, had fled from Hitler.

Yaroslavsky hastily shut the window and coughed severely. It must be the Nabob—but no, there was more than one person. I heard the Italian's door opened, and a kind of slithering sound which I could not identify. Why did it remind me of flower beds? Good God, it was a hose! So Vevers had not been joking when he had said to me: "We'll hose you down with freezing water and then shove you in a punishment cell."

The wails became shorter as the victim gasped for breath. Soon it was a tiny shrill sound, like a gnat's. The hose played again; then I heard blows being struck, and the iron door was slammed to. Dead silence.

According to my reckoning, we were now on the eve of December the fifth—Constitution Day. . . . I do not remember how I spent the rest of that night. But I can still hear the exact sound of the Italian woman's thin piping voice as I write now, a quarter of a century later. . . .

Yaroslavsky had gone off duty. The door was flung open. Before I could see who it was, I realized with brutal clarity what I myself looked like, crouching on that filthy, frozen, stone floor, disheveled and covered by a dirty smock, with enormous sandals on my blue, frostbitten feet. A hunted animal—who could wish to live after such an experience?

Apparently one could. The person in the doorway was the wardress called Dumpling. Her cheeks and chin were dimpled, her hair fell in cheerful ringlets onto her prison uniform, she smelled overpoweringly of scented soap and strong eau de Cologne. She said something in a kindly voice. At first I could only take in the sound of her speech—could there really be such a thing as a kind, pleasant, friendly human voice? Gradually I understood what she was saying.

"You're going back to your cell now. It's nearly suppertime. Tomorrow you can have a wash in the shower room."

"Supper? I thought it was morning."

She helped me to take off the smock and put on the gray-blue prison uniform again. How nice and comfortable it seemed! At once I felt warmer, and for the first time in five days I stopped shivering. I tried to put on my stockings but somehow could not—they seemed terribly long, and I could not remember how to pull them on. She helped me again and said: "Come along." We came to a space off which three or four cell doors opened. Outside each one was a pair of down-at-heel shoes—evidently there were not enough regulation boots to go around. But what was this? A pair of exquisite high-heeled shoes, size three at the outside—evidently the Italian woman's; and I thought of that slight, graceful figure, and the hose torture. . . .

Soon I was being taken up the stairs which I used to go up every day after my walk. Suddenly I stopped in agonizing uncertainty: was it to the right or to the left? I began to be frightened. I must be losing my reason after all. I had followed this route every single day for three and a half months. Why had I suddenly forgotten it?

"To the right," Dumpling whispered. I tried to turn, but now I lost all sense of direction and slowly collapsed on the stairs. Evidently I was not made of iron after all.

As I sank into unconsciousness, the last sound that rang in my head was that piercing shriek and the words *"Comunista Italiana! Comunista Italiana!"*

39

• *"Next year in Jerusalem"*

Everything on earth comes to an end, and so did the accursed year 1937. As December wore to a close we were both suffering in health as a result of our experience in the punishment cells: I had frostbitten feet and Julia's lungs had got much worse.

The newspapers were full of reports of the first elections held under the new constitution * and the new electoral law.

Day and night we were tormented by the thought: Does

---

* The so-called Stalin Constitution of 1936, hailed at the time as "the most democratic in the world."

our disappearance from life mean nothing to anyone? I imagined my ex-university colleagues voting. Did none of them remember me? And what about the people I had worked with on the newspaper? But no doubt the turnover there had been a hundred per cent.

Even now—we asked ourselves—after all that has happened to us, would we vote for any other than the Soviet system, which seemed as much a part of us as our hearts and as natural to us as breathing? Everything I had in the world—the thousands of books I had read, memories of my youth, and the very endurance which was now keeping me from going under—all this had been given me by the Soviet system, and the revolution which had transformed my world while I was still a child. How exciting life had been and how gloriously everything had begun! What in God's name had happened to us all?

"Julia! Wake up! Listen, don't you think 'he' must have gone mad? They say that megalomania and persecution mania go together. . . . Suppose, like Boris Godunov, he really does 'dream of slaughtered infants all night long'?"

Julia mumbled something unintelligible. Why had I wakened her up just to say that, especially when she had not got over her spell in the punishment cell?

Illness, however, did not prevent Julia from preparing to see the New Year in. Each day she set aside one of the two lumps of sugar we were given, and for more than a week she had been jealously saving half an ounce of butter out of the seven ounces (for three months) which we had bought from the prison store.

"We must do the thing properly. You know, there's a saying that the way you see it in makes the whole year good or bad for you. And you must write some special New Year verses."

"I wouldn't be in too much of a hurry to rejoice if I

were you. For all we know, they'll be celebrating the New Year in an even bigger way than the first of December—the anniversary of Kirov's murder. I shouldn't be surprised if they've invented a special kind of punishment cell for the occasion!"

All the same, as the New Year drew closer we both looked forward to it with impatience. We felt, superstitiously, that the fresh air of 1938 might blow away the nightmares of 1937.

As was always the case before dates of any significance, the prison regime increased in severity. More and more often we heard the click of the peephole, the malevolent hiss "Stop talking." The library for some reason was not functioning, and we knew almost by heart the books we had had for the past month. One of them was the volume of poems by Nekrasov, which Julia could not bring herself to part with. . . . I reflected once again on the power which literature exerts on us in that state of spiritual composure which prison life induces, and which makes us strive, devoutly and humbly, to drink in an author's words to the full. I have never loved human beings so devotedly as in those months and years when, cast away in the inhuman land and imprisoned behind stone walls, I absorbed every line of print as though it were a message radioed from Earth, my distant mother and homeland, where I had lived with my human brothers and sisters, and where they lived still.

Even Nekrasov's most hackneyed lines were now as moving as a letter, charged with emotion, from a distant friend. I used to read to Julia "Knight for an Hour" and "The Russian Women." Strange to say, the passages which struck one most forcibly were those which one had previously hardly noticed. For example, "Sleep is for others, not for me"—the poet's exclamation on a moonlit, frosty

night when he "longed to sob upon a distant grave." I
had learned these lines at school and they had merged
with dozens of other lines describing landscapes and the
feelings they aroused, without making any deep impression on me; but now . . . ! Outside our cell window was
that very same frosty night, and however cold it might be
we never closed the chink through which a trickle of fresh
air came in to us, and with it our beggar's dole from the
table of life. It also brought with it Nekrasov's tingling
frosty night, and I felt as if no one before us could have
fully appreciated the words: "Sleep is for others, not for
me."

Apart from Nekrasov, one of the books on our little·
table hinged to the wall was a volume of poems by Selvinsky.* In it we happened to read some lines about the des-
tiny of the Jewish people, their stubbornness and vitality,
and the ancient greeting: "Next year in Jerusalem!" I took
this as the theme of the New Year verses Julia had asked
me for:

And again, like gray-haired Jews,
We shall cry out eagerly
With voices cracked and weak:
Next year in Jerusalem!

Clinking our prison mugs
We shall drain them dry—
There is no sweeter wine
Than the wine of hope!

Comrade, be of good cheer,
Our prison food is not manna from heaven
But, like the gray-haired Jews,
We believe in the Promised Land.

* Ilya Selvinsky, b. 1899, Soviet poet.

We may be poor and persecuted,
But on New Year's Eve
We'll cry:
Next year in Jerusalem.

So at last New Year's Eve arrived—our first in prison. If we had known then that no less than seventeen of them lay before us—if, on a television screen in our cell, we had beheld even one of the scenes we were to endure in the next seventeen years—I doubt if we should have greeted this one with such fortitude. Luckily, the future was closed to us, and hope lulled us with its childish babble. In defiance of logic and common sense, we were confident that before the year was out we should be "in Jerusalem."

Lying on our prison beds, we did our best to gauge the passage of time. It was not easy—not for nothing had Vera Figner entitled her memoirs of solitary confinement "When the Clock Stopped." But during the fateful minutes when that extraordinary year was vanishing into the abyss of time, giving place to a new one which would, we believed, be its righteous judge, we were able to count the steps of time by many intangible signs: the beating of our hearts, the warder's expression as he looked through the peephole. Prompted by some sixth sense, we simultaneously stretched out our hands from under the gray, prickly blankets and clinked together our tin mugs, which we had previously filled with sugar and water. Luckily, the warders did not notice our illegal gesture, and we were able to drink the toast in peace, eating our bread with its thin spread of butter—a truly Lucullan feast!

I read my poem triumphantly to Julia, and we fell asleep dreaming sweetly of the months to come. Next year in Jerusalem!

40

• *Day after day, month after month*

But the miracle did not take place. The new year did not judge the old, but emulated and in some ways outdid it.

The year 1938, which we spent in our solitary cell from start to finish, seemed to last an eternity and yet to be over in a flash. Each day was unendurably long, but the weeks and months went by at a gallop.

At six in the morning the flap-window clicks open and gapes like the jaws of a dragon. It barks out "Get up!" and clangs shut again. In winter it is still dark, and we cannot tell where the window screen ends and the strip of sky above it begins.

"Get up" means: Realize once again where you are, and that all this is not a dream. Everything is still there: the dark-red walls, the creaking iron bunks, the stinking bucket, the niche in the wall—"Nicky"—with its cracks that may be taken for the outline of a man's face, and in which Julia fears that a microphone may lurk.

We jump out of bed—it is against the rules to dawdle—and pull on our "Yezhov" creations. Shuddering because of the damp, the stench, and our disgust with life, we try to make jokes about each other's appearance. In one of our games I say to Julia: "Would you be wanting a cook now, Ma'am?" and she replies severely: "Have you a work permit?"—"Well, no, Ma'am, I wouldn't be deceiving you, I haven't got one exactly." At first this made us laugh, and we go on with it now to keep up our courage and hide our morning despair from each other.

Sometimes it is more than an hour before we are taken out to the washroom, but we have to be ready to jump up instantly and grasp the slop bucket when the door opens. To while away the time of waiting, we may manage to sleep on the iron flap-seat or, screening each other in turn from the peephole, do "illegal" physical exercises to warm up our hands and feet, which have got numb from contact with the rough iron bunk. From vague noises in the corridor, footsteps and even breathing, we can tell which of the warders is on duty. This is very important. With St. George, for instance, we can do exercises almost openly and he will pretend not to notice. With Dumpling, we can ask for a needle and get on with mending our stockings. But if it's Worm, then look out! The minute you so much as raise your arms, he'll have the window open and croak like a toad: "That'll do! This isn't a gymnasium!" As for the Nabob—he won't say a word, but he'll report you immediately, and then good-by to walks, library books, store privileges.

Our visit to the washroom broadens our horizon, as it brings us additional political news. The squares of paper dealt out to us may contain exciting information. In the cell we get only the *Northern Worker*, but here we often see bits of *Pravda* and *Izvestia*. We read them closely, incomplete sentences and all, and draw all sorts of conclusions. . . .

Back in our cell, we wash each other over the slop bucket and breakfast on bread and hot water. Julia keeps a piece of sugar till suppertime, but I eat both in the morning, always telling myself: We might die before evening, and then what a waste. . . .

After this, our "working day" begins: reading and writing, writing and reading. Unfortunately, we often have to read the same books over and over again, as the library is always being closed for stock-taking, or something. And whatever we write has to be erased, since on the thirtieth of every month we have to turn in our two notebooks to the censor, who does not return them. I write lots of poems, learn them by heart, and rub them out with bread crumbs. I also write a novel about my school life in the days just after the revolution—it gives great pleasure to Julia, my sole reader and critic.

After each chapter we recall happy childhood memories —for in that dawn of the revolution we had such a childhood as no one has ever had before or since. Even that famous poster bearing the picture of an enormous louse, and summoning us to the fight against typhus, is now sheer poetry to us. And the establishment of self-government in the schools! And our first political parade, in which we marched with wet feet and holes in our shoes, carrying a banner with our very own slogan: "The school of hard work and joy salutes Soviet power."

The only thing we regret is that we did not cram enough activity into our lives. If we had known that our lives were to be cut off at thirty, we would have worked harder and left something to be remembered by. Instead of two children I should have had at least five, so as to leave as much of myself behind as possible. Oh, how much better we would do things now if we could begin over again!

Dinnertime. If we get *kasha* made of corn, it's a good day: I can hardly touch the "shrapnel" kind. I am thinner now than I was at fifteen. . . . Our afternoon walk. The door is flung open, and the foul smell from the bucket mingles with the warders' scent—under the regulations they are issued perfume to help them bear the stench. . . . Dressed in our hideous jackets, we pace conscientiously up and down the fifteen-yard space, trying to steal a glimpse of the sky. It is against the rules to look at it openly; the order is to keep our heads down. Then back to reading and writing, algebra sums and long conversations in which we discuss our lives from every imaginable angle.

Supper. Day after day, at the same hour, the corridor is filled with the stupefying smell of fish. Not only the soup itself, but the very smell makes me sick. I live mostly on bread and boiling water. Julia says I have turned from a plump brunette into a skinny, mousy creature, my hair having lost its color owing to the darkness we live in. I can't tell: it is over a year since I saw my face in a mirror, or even reflected in a window or in water. . . . The dragon's jaws of the flap-window clank open again: Lights out. This is welcome, this is almost happiness. We can lie down and go to sleep stretched out instead of crouching on the flap-seat. Seven hours of Nirvana, of blessed extinction. Instead of just saying "Good night" I repeat to Julia Nekrasov's lines about "Kind sleep that makes the prisoner a king. . . ."

Day after day went by in this fashion, but the monotony was likely to be broken at any moment by unexpected events. Sometimes the dead silence of the corridor was disturbed by knocking, groans, blows, and the choking screams of victims. Clearly it was not we alone who were dragged to the punishment cells, or to worse places, and some prisoners offered resistance.

Further variety was provided by searches and baths. The shower cubicle, though it barely held the two of us, was a source of great pleasure. As for the searches, they required much concentration, ingenuity, and quickness on our part. One might wonder what there was to search for in a cell whose occupants never received anything from outside and never went out except to exercise in the prison yard. But we did have things to hide: the forbidden brassieres, fishbone needles from the evening soup, and medicines from the nurse who came around from time to time. These were supposed to be swallowed in the presence of the nurse and warder, but we wanted to have some in reserve in case of an attack of malaria, for instance, from which we suffered a great deal. So we pretended to swallow the quinine powders and aspirins but actually hid them in our brassières.

We were able to conceal these illegal possessions by dint of much acrobatic skill and thanks to the fact that the cell itself was searched by men, but the search of our persons was carried out by women. The men would burst in like a whirlwind, the surprise element being clearly an important part of their instructions. They tossed the straw mattress and pillow about and scrutinized every square inch of the floor and walls. During this time we kept the forbidden articles in our stockings or brassières. The critical moment was when the men went away and the women came in. During that instant we had to stow all our contraband under the mattresses, which the women did not look at: they ran their hands over our bodies, made us open our mouths and undo our hair, spread out our fingers and toes, and so forth.

In the whole of that year there was perhaps one joyful event: at the beginning of spring we were able to get from the library a large one-volume edition of Mayakovsky's poems. We reread his early prison verses about Butyrki and the sunbeam on the wall for which, he said, he would then have given all he possessed. As for us, we were less demanding—we would have been content with a scrap of ordinary daylight, to save us from the headaches we got

from reading in perpetual darkness. We read nothing but
Mayakovsky for several weeks, and I addressed verses to
him in his style:

> Mayakovsky, you were very good
> At putting your finger on essentials . . .
> You know, here one would never think
> That such a place as Nice existed. . . .

We talked for many long evenings about the silly hero
worship of Mayakovsky which had already begun before
we went to prison. So I wrote, and then rubbed out:

> Mayakovsky, listen:
> If we live,
> We'll surely wash you clean
> Of all this fancy dirt.

> And then, elbowing the clouds aside,
> Handsome and twenty-two,
> You'll start your march
> To eternity again. . . .

41

• *A breath of oxygen*

One day we heard with alarm a series of bangs indicat-
ing that cell doors were being opened and shut one after
another. At this time we were already in an uneasy mood.
Not long before, we had been deprived of newspapers for

a month for some fancied infringement of the rules such as "talking too loud"—the Nabob and his colleagues were not a very inventive lot. When we did again receive the *Northern Worker*, it was full of the Bukharin-Rykov trial, which had apparently just begun, although at Butyrki we had thought it was over long before. . . . More of Vyshinsky's ranting speeches, and the mystifying "confessions" of the accused. We racked our brains all day to account for their behavior. Were they so afraid of death? They might have been wrong a hundred times over, but after all they were major political figures. In Tsarist times they had not been so cowardly. Were they out of their minds? But then they would have behaved like Van der Lubbe at the Leipzig trial, sitting in dazed silence and occasionally crying out "No, no!" But these men made long speeches, couched in the Bukharin style. Could it possibly be that they were impersonated by actors? After all, Gelovani could play Stalin so that no one could tell the difference.

At this time we were also depressed and profoundly shaken by the death of Lenin's widow Krupskaya. We wept bitterly as we looked at the small photograph in the local newspaper—I believe it was the first time we had cried since we arrived at Yaroslavl. The obituary was cold and matter-of-fact. The "boss," of course, did not like her: he was supposed to have said, "If you get up to any tricks, we can easily find someone else to be Lenin's widow."

We gazed at her kind, rather bulging eyes, her schoolmarm's collar, her straight, smooth, gray hair. Everything in her appearance was familiar, homely, close and dear to us. Her death was like the final act of a tragedy: the last decent, honorable figures were disappearing from the stage, dying or being destroyed.

And again the torturing question: Are any of the others like Krupskaya still free, do they know what is going on, and if so, why are they silent? Why, for instance, doesn't Postyshev speak out? Julia had known him personally and considered him a model Leninist; we did not know then that he had shared the fate of so many others. But how could he speak out, and what use would it be if he did? He would simply join all the other thousands of victims. In the present reign of terror, it was not that they cared for their own skin, but that action would be useless. The only course for such men as he was to preserve themselves for better times.

It was while we were oppressed by thoughts like these that we heard the banging sounds. Finally they reached us. A block warder—not "Pipsqueak," who brought us letters, took us to the dentist, and performed other errands of mercy, but the tall, lean, and impassive "Borzoi"—came in with a stool in his hand which he put underneath the window. He fiddled about with the fanlight, shut it with a bang, and locked it with a big iron key. We were dumfounded—so much so that we asked him a question, although we knew quite well that within these walls questions were not answered and there was no point in asking them. Anyway, it was silly to ask why when the reason was perfectly clear. It was so that we would die the sooner for lack of air; so that the walls would get even moldier and our joints suffer more from the damp. This, of course, was by way of a "response" to the Bukharin trial. The principle was a familiar one: it reminded us of the "rubber-stamp" resolutions described by Ilf and Petrov, which all began "In reply to . . ." or "In response to . . . ," and where you filled in on the dotted line "the machinations of the Entente" or "increased output by workers in the public utilities." This time, it was "in response to the trial of the right wing." It certainly showed that some very ingenious brain was at work.

As Borzoi left us, he mumbled, "It'll be opened for ten minutes a day." Now we really got to know the taste of air—our tiny ration of oxygen . . . The ten minutes were supposed to coincide with our exercise period, but if the kindly Yaroslavsky or St. George was on duty he would open it not when we were led out of the cell but after the order to get ready, so that we had an extra five minutes of fresh air which we greedily drank in as it came through the small square pane, so high up that not even the lanky Borzoi could reach it without standing on a stool. The days and nights when it had been open all the time were, in comparison to the present, like days spent at a sanatorium. After a short spell of the new oxygen ration, the damp in our cell, which faced north and never received a gleam of sunshine, became unendurable. Any bread we kept till dinnertime became moldy. The walls were green all over, our underclothes were always damp, and all our joints ached as though some creature were gnawing at them.

At this time I kept dreaming that I was sitting on the terrace of a country villa on the opposite bank of the Volga from Kazan. The awning over the terrace swelled like a sail with the gusts that blew from the river, and I breathed for all I was worth. But for some reason I felt no better, and I had stabbing pains at the heart. . . . Then the warder would bark "Get up!," I would open my eyes and see the locked pane. The crows perched on the wooden screen looked through it inquisitively, holding their long beaks sideways. . . .

42

"Why are you coughing like that?" asked Julia.

"Well, so are you."

"Yes, but I've got pleurisy."

I had realized for a long while that the sharp tickling in my throat was connected with the smell of burning which was increasingly noticeable in the cell. But I had said nothing so as not to frighten Julia, who had looked gray and sallow ever since she had come back from the punishment cell. Anyway, perhaps it was nothing serious, something burned in the kitchen, for instance. But as far as we knew there was no kitchen in this block; the food was brought in from outside. We were coughing more and more, but went on reading until our eyes started to stream and glazed over. Then we heard feet running about on the roof and water gushing out of hoses. Other people were rushing along the corridor and even talking in loud whispers. After a time we heard a faint knocking on the wall: it was Olga Orlovskaya tapping out the word which Julia and I had been afraid to say to each other: F-i-r-e . . .

"They'll have to let us out," I said in reply to the question in Julia's wide-open eyes. "It can't be part of their plan to let the prisoners suffocate in their cells, at least not all of them at once."

In a few minutes the cell was so full of black, acrid smoke that we could hardly breathe. "I'm going to ring," said Julia firmly. "The swine could at least open the window!" And she pressed the silent button which we were allowed to use in emergencies only, and which operated a light signal on the desk of the warder on duty. After a while the flap-window crashed open and the thin-lipped, pimply Worm put his head through. "What is it?" he asked in a surly whisper.

"Can't you please open the outer window? We're suffocating!" begged Julia.

He slammed down the iron flap, almost hitting Julia in the face, and replied through the closed door: "It'll be opened if necessary." Meanwhile the panic was increasing. The clatter of soldiers' boots on the roof was getting louder, and from the corridor we heard not only whispering but confused shouts. The cells, too, broke their deathly stillness. Some of the prisoners, evidently despairing of attracting attention by means of the bell, started to hammer on their doors. Olga was signaling almost openly: the guards had other things to do than to listen to us. She tapped out: "Looks as if . . . they want us . . . to suffocate . . . to death."

"The sugar!" Julia shouted, wringing her hands. We had drawn a pound of it from the store on the previous day. "I'm damned if they're going to get any!" she added, without the least suggestion of a joke. "Come on, let's eat it all up!" And we started to eat it in handfuls—far from being cloying in such quantity, it seemed to us food for the gods. In between, I bit off mouthfuls from a hunk of bread. We ate the whole pound of precious sugar, grinding our teeth fiercely and pausing to cough out the smoke, which was now so thick that we could not see each other.

"Let's sit together, Genia," said Julia in tears, "and say good-by." We kissed and hugged each other, and then, in defiance of all the rules, sat side by side on Julia's bed with our arms about each other's shoulders. I saw with horror

that her round, slightly unsymmetrical eyes were beginning to bulge: her face was turning blue and the veins were
standing out like whipcord. Pray God she would not die
first. . . .

By now the whole place was alive with the noise of
prisoners hammering on their cell doors and shouting:
"Open up! We're suffocating! You've no right! Open!"
Colored sparks were dancing in front of my eyes: I could
not tell if they were real ones from the fire or whether
I was losing consciousness. Then, amid the hubbub in the
corridor I heard the sound of keys turning in one cell door
after another. I shook Julia by the shoulder.

"They're letting us out! Julia, just hold out for a little
bit longer. Do you understand? They're letting us out!"
The smoke was getting blacker; Julia was choking in my
arms. Could I knock out the pane? It didn't matter what
we did now. I stood on the bed, but couldn't reach it.
This must be the end; how frightful and unexpected it
was! In our prison conversations we had imagined many
forms of death, but never death by fire. . . .

"Out you come!" Our door was flung open and Worm,
dressed in a grimy tunic, panting and sweating, dragged
out the fainting Julia more or less by the scruff of her
neck. I said: "I can manage by myself," and he ordered
me downstairs. . . .

Yes, Worm and his fellow thugs were professionals, all
right. Even in that panic they succeeded in not breaking
the rules of solitary confinement. Where they put all the
prisoners I don't know to this day, but still the fact is that
Julia and I were taken out into the exercise yard by ourselves: we met no one else and saw no one else. . . .
Nevertheless, the ill wind of that day brought some good
—we were allowed to breathe freely out of doors for no

less than an hour and a half. Julia, who felt better by now, winked knowingly at me and looked up at the sky, as though to say: "How clever of us to get this extra walk!" Next day Olga informed us through the wall that she, on her walk, had seen absolutely no one either.

43

• *Punishment cell for the second time*

At the end of June 1938 I received a letter from Mother which said: "Darling Genia, your father died on May 31st. When you think of his life and all the things he did, all the children and grandchildren he had, and there were only two people at his funeral—Claudia the washer-woman, and I." Exactly half an hour after I received this letter, the door opened again and the Nabob ordered me to follow him. . . .

Doubtless death itself would not be so frightful if one endured it a second time. . . . I walked down and down with assurance, knowing where I was going, without the same horror as in December—in fact, with a sort of numbed indifference. I imagine that in this state it would not be difficult to stand up and face a firing squad.

Yes, the same cell; the same ragged smock and bast sandals, the same bed of planks an inch or two from the floor, the same inferno. But I was not afraid this time, I did not cry out or resist. I listened almost with indifference to the Nabob as he read out the order: three days for breaking prison regulations by singing in my cell. I didn't bother to retort that there had been no singing by anyone at any time: what would have been the use?

The door banged and I was alone in the darkness—alone with thoughts of my father's death. It was summer. One sign of this was the fact that, besides rats, the place was crawling with all sorts of vermin: beetles, roaches, centipedes. . . . By now I felt like a veteran of the punishment block: I could keep count of time and was used to sleeping on two boards, but I was still determined not to eat anything. This, I said to myself, was a good sign: I still possessed the human attribute of fastidiousness. I did not know then that I had before me years of camp life during which I and my fellow prisoners periodically lost this attribute and would eat any filth in order to keep alive.

Groping around in the dark, I identified each part of the well-known punishment-cell equipment: the smock made from military cloth, the sandals, the rusty tin mug on the floor. But somehow I felt so devastated that I couldn't take these surroundings as tragically as I had the time before, in winter. I did not wish to scream or struggle. . . . Now I think that, in Germany too, no doubt, they got used to the gallows and the gas chamber. People can get used to anything. I found myself thinking that it was a good thing I had only three days to go, not five, and also that I was only in a "lower" cell, which was the second of three categories. I had learned about these during the past few months from Olga, who had the information from her neighbor on the other side. In the first class of cell, which was less severe than mine, there was a lamp burning and one did not have to remove one's ordinary prison clothing. But there was also a third kind, which thank heaven I never experienced, from which one emerged only in a half-dying condition. . . . The class of cell that one was assigned to did not, it appeared, depend on the enormity of one's pretended offense, but on the color of the stripe on one's personal file: Julia and I were both "second class."

I spent most of the three days standing up. I accustomed myself to standing on the boards, as far away as possible from the stone, covered with a kind of slimy, grayish hoarfrost, which was supposed to serve as a pillow. I stood for hours till my strength gave way and I sank into a short, oppressive sleep, from which I was usually awakened by the pain and itching in my frostbitten toes. The pain brought everything back to me: I was still I, this was still going on. I stood on the planks again and could see Father as if he were still alive. I could not picture him as a dead body. How little I really knew this man who had given me life—and yet, at the same time, how indissoluble was the tie of blood! I felt as if everything within me was a single great lump of pain. Father—*my* father. It was as if a part of me had died.

It was well for me that I did not know then the circumstances of his death, about which they wrote me years later in Kolyma. I did not even know that my parents had been under arrest—for two months only, but it was enough to kill my father. When they came out of prison their apartment was occupied by strangers and their things had been confiscated. They wandered, in search of a night's lodging, through the town where Father had been an honest and respected technician, where his daughter and son-in-law had worked as Party members. Everyone was scared stiff of the old couple, no one would take them in: only Claudia, the washer-woman, showed some decency. All this had happened while I was at Butyrki, but I did not hear about it until three years later.

As I went on thinking about Father, tender scenes from my early childhood began to flit across the walls of my dungeon. What fun it was to play with his watch chain as I sat on his knee! Then, when he took me out for walks, he used to teach me strange Greek words and tell me about his schooldays. Born in the nineteenth century, he had learned both Latin and Greek at high school. . . . Up to my eighth year we were as close as two human beings could be. Then there were long years of argument, antagonism, recrimination. I was irked by our "social background" and envied those of my friends whose fathers were "true proletarians"; he disapproved of many things in my life and conduct. . . . Suddenly I saw before me the words of the last letter I had received from him here at Yaroslavl: "I must admit that I haven't felt up to much lately; but I will do my best to stay alive for the sake of my dear grandsons, Alyosha and Vasya." For their sakes! . . . And I should never be able to ask his forgiveness.

As I stood there, I hoped that fatigue would soon overcome me and that I should again fall into the state of drowsiness that passed for sleep. Then the three days would be over more quickly. As before, I refused to eat any bread, but this time no one came to tell me that hunger strikes were forbidden. Apparently they too could get used to things. Or perhaps what they wanted most was for me to sit out my time quietly. If I did not eat, I was more likely to fall into a comalike state of exhaustion.

When I was here in winter I had kept thinking about the outside world, and wondering if people there knew about these dungeons and tortures. But now I could hardly believe that the outside world existed. How could I, for instance, believe that it was summertime and that at this moment people were bathing in the river?

I composed some verses about my second spell in the punishment cell:

It's like the Holy Inquisition,
My bare feet stand on icy stone—
Am I accused of dealing with the devil
Or of opposing the Party line?

The centuries merge and blur
In this renovated dungeon.
Is that Princess Tarakanova
Breathing her last here at my side? . . .

44

• *Memories of Giordano Bruno*

The summer of 1938 was fearfully hot in Yaroslavl. The *Northern Worker* informed us of this fact day after day: there were graphic descriptions of melting asphalt and figures of average temperatures for previous summers showing that this was the "hottest ever." But our cell window remained shut, and all our things became musty from damp mold and stale air. The straw in our mattresses and pillows rotted and began to smell.

After our second spell in the punishment block we both fell ill in good earnest. We could no longer swallow the bread and thin soup. I had had to ask for a needle on three successive occasions in order to alter the position of the hooks on my prison skirt. Julia's face was a livid color, with yellow circles under the eyes, and as I looked at her I felt that neither of us would last much longer. To make matters worse, my bouts of malaria had returned, no doubt owing to the stifling dampness of the cell. After these fits my heart refused to work properly, and on one occasion I fainted. Julia pressed the bell and asked for a doctor. I must have looked like a corpse, for although the warder on duty was Worm he summoned the doctor without a word.

This was our first contact with the prison medical system except for the periodic visits of the nurse with her box of medicines, which included aspirin "for headaches," quinine for malaria, and salol "for stomach-aches," plus the inevitable valerian for everything else. When I regained consciousness I saw the doctor's face bending over me and was struck by its humanity. It was a true doctor's face: attentive, wise, and kind. It seemed to take one back to the world outside prison bars, tormenting one with hundreds of memories. . . . Remembering his round, soft features and the kindness plainly seen in every wrinkle, we decided to call him Dr. Pickwick, feeling that his fellow students must surely have done likewise.

"There you are, then," he muttered as he drew the syringe out of my arm, which was as thin as a matchstick. "You'll feel better when the camphor starts working, and Nurse'll bring you some medicine twice a day."

"But doesn't she need more than medicine?" The bold question came from Julia, who was scared to death at the prospect of losing me. "I think she's starved of oxygen, especially as it's so hot outside. Couldn't you give orders for the windowpane to be left open, Doctor, seeing how terribly ill she is?"

Dr. Pickwick's face slowly flushed brick-red. He looked out of the corner of his eye at Pipsqueak, the block warder standing behind him (doctors were not allowed to enter the cells alone), and said in a low voice: "That is outside my competence."

Pipsqueak coughed by way of confirmation and added: "Conversation must be confined to the subject of the illness."

During the agonizing days that followed, the flicker of life within me was kept alive only by my inexhaustible curiosity. I must see things through to the end—even if it were my own end.

But, in spite of my optimism, I began to notice several dangerous symptoms in myself. For instance, I refused several times to take my afternoon walk, and when Julia tried in a frightened whisper to persuade me not to forfeit the little fresh air I was entitled to, I answered wearily that I wouldn't be able to climb the two floors back again. . . . Nor, indeed, was it possible to walk up and down the fifteen-yard-long outdoor cell when one's heart refused to supply the necessary energy.

Making jokes also became more difficult day by day. None the less, we still occasionally resorted to that well-tried cure for all ills. Our favorite joke was the story of the incorrigible optimist: "Oh, well, if it's a common grave, that's some consolation." And, when the cell was more asphyxiating than usual: "Giordano Bruno was worse off—his cell was made of lead!"

After the doctor's visit, we had a long argument over what one should think about a doctor's working in prison.

"I bet he'll dream about you all night. He's a good fellow, our Pickwick."

"Of course he is. All the more shame on him that he does a job like that."

"Well, would you rather it were some 'Nabob' with a doctor's degree?"

Events showed that Julia was right: within two days we had proof of what Pickwick could do for us.

"Get ready for your walk!"

"I can't go—I'm not strong enough."

"Yes, you can. They've put a stool there. You're to sit outside for fifteen minutes—doctor's orders."

And we really felt affectionate toward Pickwick when the wardress called Dumpling, as she unlocked the pane in our window with an enormous key looking like a stage prop, whispered encouragingly:

"Doctor says you're to have the window open twenty minutes a day instead of ten."

Although there was not a breath of wind to stir the stifling air, we looked at each other overjoyed.

"There, you see! Giordano Bruno's cell was made of lead!"

45

• *The end of the "monstrous dwarf"*

As far as we were concerned, the whole period from summer 1938 to spring 1939 might be labeled, in Vera Figner's phrase, as one "when the clock stopped." Our acute physical sufferings, the constant lack of air, the body's struggle for each breath it drew—all this combined to cloud our awareness of the passage of time and of the outside world. Today, those months appear to me as a kind of black stream, continuously flowing yet motionless. The clocks of our lives had stopped, and nothing could set them in motion—certainly not the pale reflections of a distant, alien life which were purveyed to us daily by that semi-literate, brash, yet incredibly dismal journal, the *Northern Worker.* Certainly, from force of habit we devoured it as greedily as ever, examining every line more carefully than any proofreader; but nothing we read in it seemed wholly real.

The Spanish war—Munich—Hitler in Czechoslovakia—preparations for the Seventeenth Party Congress . . . Surely all this was some kind of dream! Was anybody putting up a fight anywhere in the world? Was not everyone as broken in spirit as we were? . . . Day by day we fell deeper into that dangerous state, a sure sign of the end, in which one feels detached from everything that is taking place outside one.

Even inwardly, I believe, we had given up hating or protesting. I looked back with amazement at the time in December 1937 when I was first taken to the punishment cell and hit at the Nabob with my fists. Such an explosion of physical and spiritual energy now seemed to me unthinkable.

On the eve of 1939, which had stolen upon us imperceptibly, I composed another poem at Julia's request, but it was not in the optimistic vein of "next year in Jerusalem." This time I expressed my fears that we might get so used to our prison life that the outside world would be meaningless to us if ever we were released.

Our hopes had dried up, not least because of our physical exhaustion, which increased daily.

"Genia! Open your eyes! Can you hear me? Open them, or at least move your hand." When Julia badgered me in this way I would reply: "Yes, I know I look like a corpse. Don't worry, I'm alive. You don't look so marvelous either!"

Julia's cheekbones, the skin drawn tightly over them, stood out sharply; there were deathlike circles under her eyes. We were lucky not to have seen our own reflections in a looking glass for over two years; but each of us, by looking at the other, could imagine how close she herself was to the inevitable end.

We had completely lost our appetites. Every morning we used to drop almost all our bread in the waste basket at the end of the corridor. We felt at times that only the two daily lumps of sugar somehow sustained our ghostly existence.

In March we were once again punished, this time for "marking library books with our fingernails." They evidently realized we were in no condition to be put in the punishment block, but the prison had its own quota of punishment to fulfill. So, on top of everything else, we were deprived of newspapers for a month. Our only link with the outside was thus broken except for the rare occasions when we heard Olga's faint knocking through the dark-red wall. . . .

"The Par-ty Con-gress has be-gun."

I had the energy to reply sardonically: "The survivors' congress?" She went on:

"They men-tion ex-cess-es and re-ha-bil-i-ta-tions."

Julia was at once full of hope. "Do you hear, Genia—rehabilitations! At last! Well, it had to come. . . ."

But I, the pessimist of our company, could not believe that the rehabilitations would be on a large scale, or less random than the arrests. I had got to know Stalin's style too well during the past two and a half years. Nor had I any hope that I personally would draw a lucky number in the new lottery that was just beginning.

Suddenly . . .

How often have I used that word! I realize that it makes my story monotonous, but it can't be helped: the abruptness is in the facts themselves. Every now and then our jailors would burst into the dank, moldy, stagnant cell to remind us that, unlike run-of-the-mill corpses, we could still be tortured physically and morally, subtly or crudely, together and separately, by night and by day.

Suddenly, then, we heard the noise of cell doors being opened and shut one after the other, which was always the sign that something new was afoot. In due course Borzoi came in, the same who had locked our outside windowpane. His face, which reminded us strongly of an elderly dog of the breed after which we had nicknamed him, today wore a curious expression which almost suggested embarrassment, though surely no one could stay long in this job who was capable of such a feeling. Without a word he stretched out his hand to the right-hand wall and—were we dreaming?—took down the piece of cardboard on which the prison regulations were pasted—Major Weinstock's twenty-two commandments. Having done so, he turned sharply on his heel and went out, his polished boots shining. Then we heard him unlocking Olga's cell. . . .

This time we did not play our usual game of pessimist versus optimist: we were both certain that this boded disaster in the shape of even more stringent regulations. Perhaps it had come to Stalin's ears that we were allowed, on our deathbeds, to read books and to eat two lumps of sugar a day. Tomorrow, no doubt, there would be new commandments abolishing our right to receive books, newspapers, and correspondence, to say nothing of the prison store and the afternoon walk. . . . For a long time we had vied with our jailors in imagining things that they could still take away from us. Or again, they might have invented some new punishments. . . .

Our walk that day did nothing to cheer us up, especially as there was something strange about the warders' looks and behavior.

Several gloomy and anxious hours passed. Then we heard a commotion in the corridor once again, and a warder appeared—not Borzoi this time, but Pipsqueak. He reaffixed the commandments on the wall and walked briskly out again. His face wore the vestige of a smile, in which there was also a trace of embarrassment.

We studied the new text letter by letter, but could find no change. What on earth was all the fuss about? But—what was this? . . . We stared at each other, hardly able to believe our eyes.

"Julia, pinch me! Are we dreaming, or something?"

In the top left-hand corner of the regulations it had said: "Approved. Yezhov, Commissar-General for State Security." Well, hadn't it? . . . But now these words had been pasted over with white paper, so neatly that one could hardly notice.

For the next few hours we were at fever pitch. Had the monstrous dwarf really fallen? During the last few years the cult of Yezhov had reached fantastic proportions, so that it almost rivaled that of Stalin. He was officially styled "The nation's favorite son." All the sufferings inflicted on millions of innocent people in camps and prisons were

popularly known as "Yezhov's prickles." * Central Asian
akyns had addressed him in innumerable odes as "Father
Yezhov." Could it really be true?

We did not sleep a wink all night. Even now, a faint
hope for the future was kindled in our worn-out hearts.

In the morning Olga, who received newspapers, tapped
out to us:

"He's through. It's the end of him. He'll go the same
way as us."

We quoted gleefully: "The Moor has done his work, the
Moor can go." Then we paced about our cage like tigresses
until Olga tapped out the new man's name: "Be-ri-a. Beria."

I was almost sick with excitement, and again could find
relief only in verse.

The Monstrous Dwarf

What shall I call you? Fouché, Nero, Quilp,
Caligula, dwarf Mime, Torquemada?
Monster of many heads and stunted growth
And false as all the tales of Scheherazade!

Behold the fate that lurks for such as you,
Behold the pit wherein you now must crawl!
Yes, the Tarpeian rock, you evil crew,
Is still the nearest to the Capitol!

"That will do for the classical vein," I said to Julia.
"Now let's have an epigram in the Pushkin manner. What
do we know about him, anyway?"

We knew nothing. So I wrote:

Tell us, O Beria,
Will life be merrier,

---

* The name "Yezhov" derives from the Russian for "hedgehog."

Lighter and airier
Or even scarier?
Answer us, Beria!

Speak, speak, O Beria!
Should I be warier,
Of hope yet charier,
Outcast or pariah
For your sake, Beria?
Answer, O Beria!

From that time on, the stopped clocks of our lives quivered into motion again. They faltered and creaked a little, but the hands turned. At last we again had something to wait for. . . .

Outwardly, nothing had changed. The inflexible prison rules, the mold on the walls, the stinking fish soup and slop bucket. But in the atmosphere, in subtle details of the warders' behavior and tones of voice, we could sense the approach of great changes.

As for the commandments, they were taken away and rehung twice more in the next month or so. First the name Weinstock was painted over and Antonov substituted; then Antonov went, and in his place it read: Chief Prison Administration. "That'll save them from changing it again." We laughed—softly, but more lightheartedly than we had done since they locked up the outer window.

Thus, in full accordance with Marxist theory, the business of replacing the rules began as a tragedy and was then twice repeated as a farce.

As Julia said one evening, her gentle eyes flashing: "This'll give Major Weinstock a chance to try living by his own commandments." I myself could not sleep for imagining Yezhov, Weinstock, Antonov, and others in the lower cell. Although, of course, that was only the second degree, and no doubt they would be given a spell in the third also. "With what measure ye mete . . ."

Once on a windy, unsettling day in May we came back from our walk to find that the pane in the outer window was still open. Had they forgotten? Dumpling was on duty, and the dear creature might have left it open on purpose. But next day the same thing happened. The fall of the monstrous dwarf had restored to us our breath of oxygen. A liberal breeze was blowing in the spring and summer of 1939. Every day the *Northern Worker* printed articles denouncing slanderers. . . .

46

• *Great expectations*

Our appetites improved, we were able to eat our wretched soup, I took up my clandestine physical exercises again. But we did not read much now, because we were too busy talking.

Day and night we exchanged ideas, conjectures, fantasies —always maintaining our respective roles of optimist and pessimist. Julia was convinced that everyone's case would now be reviewed, and that we would either be set free or, at worst, deported to some place where we could live with our husbands.

"If they haven't been shot," I put in.

"But surely they weren't high up enough to be shot," she said.

Julia spent hours picturing a new happy life for us—perhaps on Dikson Island in the Kara Sea, or at Igarka on the Yenisey. We would be able to work at the jobs we had been trained for. . . .

"Teaching, for instance?"

"Why not? And we'll be reinstated in the Party, you'll see. After all, Yezhov's been unmasked, and it was all his dirty work. . . ."

The comments that Olga tapped out on the wall were a good deal more realistic. No doubt it had become uneconomical to keep so many thousands, and for all one knew millions, of people out of work. Think of the expense of the security apparatus itself! And most of the prisoners were of working age, between twenty-five and fifty. Instead of ordinary deportation, we'd probably be sent to distant hard-labor camps.

Well, even that would be almost happiness! Being moved to a camp meant traveling to new lands, rugged ones perhaps, but there would be fresh air, wind, and sometimes even sunshine, cold though it might be. And there would be people—hundreds of new people among whom we would live and work. Many of them would be good, interesting people and we would make friends with them. In short, the camp meant life—a nightmare life perhaps, but not the living grave in which we were now spending our third spring. . . . Thus we sought distraction from the pain and anxiety that were consuming us in discussions and speculations about the fate that lay ahead.

Our premonition of important changes was confirmed by the behavior of our guards, who doubtless knew for certain that we were soon to be moved from the prison.

The most callous ones, like Worm and the Nabob, did their best to keep up the "Yaroslavl spirit" until the very last day, as for example in the incident of the little flower. . . .

One day late in spring, as we were about to begin our walk, we noticed a sickly blossom which had made its way up between two slabs of asphalt. Julia saw it first and drew my attention to it with her eyes. I had paused for a second to admire this wonder of nature when Worm, who was guarding us, saw what was going on and viciously stamped the flower into the ground with his boot. Not content with that, he bent down, picked up the trampled stalk, crushed it in his fingers, and flung it to one side. . . . Throughout the walk I struggled with sobs that I could not repress. Julia, the incorrigible optimist, tried to console me by saying: "It's because we're leaving soon, that's why they are so vicious."

The first person to translate our fantasies into real terms was the friendly storekeeper, who, once a fortnight, brought around a printed form headed "Order from Prison Store," which said "Having . . . rubles in my personal account, I request permission to order . . ." Even after my father's death, my mother had continued to send me the fifty rubles a month that we were allowed to receive, so that we could buy soap, toothpaste, notebooks, Bakelite pencils, and sometimes sugar. A day or two after we filled in the form, the storekeeper would bring us such of the articles as were in stock. From the beginning I had felt that he was of a kindly disposition, though his face wore the usual impenetrable mask. Now that I had been here almost two years, there was a tacit link of friendship between us—indeed, not always even tacit: we already had our little secrets. For instance, during an illness, when the windows were closed and it was fearfully hot, I had asked in a low voice as I handed him my list:

"Could I possibly have some candy instead of sugar? Just a quarter of a pound of the cheapest kind? It's so long since I've had any."

"It's against orders," he replied.

But when, two days later, he handed me through the flap-window a tin bowl containing the things I had asked for, I found some caramels at the bottom underneath the sugar.

"Thank you," I whispered in a conspiratorial tone as I tipped the things into our own bowl, returning the empty one to him.

"Mind, you must eat them all right away and not leave any in the cell when you go for your walk."

I gave him a friendly smile and said, "Don't worry, we won't give you away." From that day we were friends. He was even closer to us than Dumpling, St. George, and Yaroslavsky: it gave them no pleasure to hurt us, but this one actually wanted to help us.

Another time, when we were not allowed newspapers, sensing as only prisoners can that there was no warder around, I ventured to ask him: "What's going on in the world?" By Yaroslavl standards this was such an impertinence, such an unheard-of crime, that I felt terrified by my own words and was even more surprised than relieved when he replied: "Looks like it'll soon be over in Spain."

So now at the end of May 1939 it was this secret well-wisher who brought us good news. I had written on the order form "Two notebooks," and he, after looking around cautiously, muttered: "You won't need them. You're being taken away soon."

In this way our dream became tangible reality: our departure was no longer a premonition or a conjecture. It is curious that at that stage of our cruel road we felt little or no interest in knowing where we were to be sent. Even if we had to walk in irons all the way to Siberia as in the old days—at all events we should be away from these walls.

We could not wait for a chance to tell Olga our news; but on hearing it, she was not surprised.

"I thought so. I was called out for a medical." So were we, two days later. The clinic turned out to be on the ground floor, where we had unwittingly passed it every day we went for a walk. There were three doctors and the inevitable block warder. They weighed and measured us and felt our muscles.

"Very emaciated," mumbled one of the doctors with a look at my ribs.

"No, no, I feel perfectly well, I can do any outdoor work." I was frozen with terror at the thought that I might be left behind. . . .

A few days later, at twilight, the warder Pipsqueak came with the familiar order "Follow me." This time there was such a cheerful expression on his round face with its pursed, fishlike lips that the idea of another spell in the punishment cell did not enter my head. We went along the corridor to the far end, where I had never been before, turned to the right, and came to a door with the notice "Prison Governor." I gazed at it, in wonder: a painted wooden door without a peephole—and locking, of course, on the inside!

Vulturidze sat at his desk, which was covered with our personal files, under a big portrait of Stalin. His first words were:

"Sit down."

Truly, a new era must have begun if he could force

such words out of himself. But it turned out that I had to sit down in order to sign a pile of papers which he shoved before me one after the other.

"Your sentence has been revised," he said through his long, nicotine-stained teeth. "Instead of ten years' imprisonment, you're to serve out your time in a corrective labor camp. Sign here to acknowledge that you've been informed."

Instead of reading the decree, I gazed spellbound at the fingernail with which he was marking the spot. How incredible that such a monster should have such an elegant oval nail! It even looked as though it had been manicured.

"You'll be leaving here," he went on.

"Where for?" I asked involuntarily, fool that I was. I had told myself a hundred times not to ask questions if I did not want fresh humiliations.

"Your personal things and the money left in your account will go with you," he said, ignoring my question. "Sign here." He marked the place with his nail again. . . .

I returned to the cell, and Pipsqueak led off Julia, who went through the same ritual. The clicking of locks and sounds of steps along the corridor went on all evening and part of the night. The prisoners were being taken to Vulturidze one by one, in strict isolation—but by now this was only force of habit. The regime of solitary confinement was crumbling, it was near its end. This was obvious from the loud voices of the prisoners in the corridor and from the warders' faces, which showed clearly that their thoughts were elsewhere—worried, no doubt, by the forthcoming reduction in staff. Even the crows perching outside looked less self-absorbed than usual.

So our Schlüsselburg period was drawing to a close. Pasternak's lines again drummed in my head their message of hope and fear. "Penal servitude—what bliss!"

47

• *A bathhouse! Just an ordinary bathhouse!*

There was nothing new about being ordered to the bath-house—it happened twice a week—but why were we told to take our quilted coats this time? What were we sup-posed to do with them in the cramped shower cubicles which only just held two of us without clothes? We discussed the order in fear and excitement. By applying Holmesian methods of deduction we came to the conclu-sion that everyone was being ordered to the baths at once and that our coats were to be disinfected. The last rite before our journey, the final act of the Schlüsselburg re-gime.

Our deduction proved to be right. As we were let out into the corridor—what an amazing, unforgettable mo-ment—the doors of the two cells next to ours were wide open, and in the half-light we saw outside them four dark female silhouettes in "Yezhov" dresses, the same as ours. How pitiful they looked, how exhausted and soured by grief—exactly as we did ourselves. Their emaciated bodies had lost their feminine roundness, and their eyes had an expression we knew too well.

The prisoners were no longer being hidden and isolated from one another: on the contrary, we were all ordered to form up in pairs. With a feeling of devotion and love I stealthily pressed the hand of the one nearest me. Then a smallish, close-cropped woman who had been standing to one side came up to me and said: "You're Genia, aren't you? I'm Olga—your special correspondent!"

"So that's what you look like," I said, smiling at her. "I pictured you quite differently through the wall."

There are no words to describe the feelings of a prisoner who for two years has seen no one but warders and suddenly comes face to face with fellow sufferers. So these were my dear friends whom I had thought I should never see! What a joy to be with them, to be able to love and help them!

We were taken out into a part of the yard I had never seen before, where about a dozen other women were standing. The warders growled "No talking!" but we were already conversing by looks, smiles, and just by squeezing each other's hands. I looked to see if I knew anyone from Butyrki, but in vain. All the faces looked more or less alike owing to the hideous uniform and the expression in every pair of eyes.

Yes, it was an ordinary bathhouse, with wooden benches steaming from the boiling water that had been poured over them, foaming soapsuds in tin basins, and the resonant echo of voices. But all these everyday details were lost to view in the exaltation of our shared humanity which possessed us at that moment.

In a way that is all too rare in life, we in that bathhouse felt true love for one another. We were not yet affected by the corrosive jungle law of the camps, which in later years—it is no use trying to hide the fact—degraded more than one of us. . . . At present, purified by our sufferings and full of the joy of meeting other human beings after two years of solitude, we felt like sisters in the highest sense of the word.

Each of us did everything she could to help the others, down to sharing her last possession with them. The small bits of prison soap were thrown away, and we used in turn a huge "family" piece handed around by a woman called Ida Yaroshevskaya. She had been given hers by the friendly storekeeper a long time before, but for some reason had kept it as if she knew that it would be needed by all.

I was given a pair of stockings by Nina Gviniashvili, an actress of the Rustaveli Theater at Tiflis. "Take them, take them! I've got two pairs, and you've practically nothing," she exclaimed as she looked at my one pair, darned with fishbone and threads of many colors. "Go on, don't be shy, it's not as if I were a stranger!"

No indeed, she wasn't. Who could say that this was the first time in my life that I had seen these enormous friendly eyes which flashed like two black diamonds?

"Thank you, Nina darling!"

We chattered all at once about everything that had happened to us. Almost all of us had been in the punishment cells, and always on special political anniversaries: the October revolution, the First of May, the First of December. . . . On the day of the fire some of the others had, it now appeared, been allowed to take their walk in twos. . . . I heard the names of Galina Serebryakova, Carola, Uglanova.*

We were so excited that none of us could finish dressing, and the wardress in charge of the bathhouse shouted herself hoarse:

"That's enough, women! Get dressed, women!"

"Women!"—this is how we were to be addressed for a long time to come.

That night we could not sleep a wink. We were glad

---

* Galina Serebryakova, wife of the old Bolshevik Leonid Serebryakov. Both were arrested during the purges. Galina S. is still alive and is known for her trilogy on the young Marx.—Uglanova, wife of Nikolay Uglanov, old Bolshevik shot in 1937.

to be with our fellow human beings again, but we were so unaccustomed to such contact that it made us quite exhausted. And somewhere deep down in us lay a wish that we did not avow even to ourselves—namely, that the break in our lives should be postponed for a day or two beyond tomorrow. Could we not live just a little longer in our cell, our very own solitary cell? We had left part of our lives in it; we had rejoiced in books and told stories of our childhood. How could we leave it all of a sudden like this? We needed time to get used to the idea, to wean ourselves from the solitary life.

"I've got a headache," complained Julia.

If we had stayed there another year or two and had not died of it, I believe we should have followed the example of Jack London's character who, on coming out of prison, built himself a lonely mountain cabin to end his days in.

"You know, Julia, there's a poem by Sasha Chorny * that says:

As a hermit poet
I should not repine
If my readers brought me
Meat, bread, and wine.

It would be rather nice, wouldn't it?"

We were again laughing in our excitement. The flap-window opened and the Nabob bellowed: "Get to sleep!"

True to the end. No doubt he was afraid of being out of a job. But, alas, it was fairly certain that he would have plenty to do for a long time to come—if not here, then somewhere else.

* Russian poet, 1880–1932.

48

• *The ruins of Schlüsselburg*

What a good thing for us that in modern times all processes have been speeded up. Those who devised and carried out the operations of 1937 found that it was simply not practical to keep such multitudes in prison for ten or twenty years: it was inconsistent with the tempo of the age and with its economy. Things were moving about ten times as fast as in the old days; and thus, instead of Vera Figner's twenty years in solitary, I served two.

At this point of time there was a "qualitative leap" in our wretched existence. Everything that had hitherto been strictly forbidden was suddenly enjoined with the same severity. For instance, at Yaroslavl there was no greater crime than to try to enter into relations with the other prisoners. This was in fact the ground on which people were condemned to the punishment block and other measures. But from the day in July 1939 on which we crossed the threshold of our cells for the last time, we were obliged to do everything together—whether it was working, sleeping, eating, bathing, or going to the lavatory. For long, long years to come, none of us could dream of being alone with herself for one moment.

"Form groups of five there! Get in line! You in back, move up! Get in step! Not so fast in front." Such commands as these, shouted hoarsely or shrilly, with anger or with slighting indifference, took the place of the warders' barely audible whispers and stealthy steps on carpeted floors.

"Just think what a lot of time and money is wasted on keeping up strict isolation," sighed the economy-minded Julia. "It used to take five warders to give one of us an afternoon walk—and now . . ."

She said this as we were looking at the Yaroslavl prisoners, several hundreds in number, who had been milling about since dawn in the exercise area. We had been assembled there for the purpose of fingerprinting and a thorough personal search—the way they did it in Butyrki—during which they made us give up the photographs of our children that we had been allowed to keep in our cells. A large number of ordinary warders, all the senior ones, and Vulturidze himself were run off their feet with work. The main object of their frenzied activity was to prevent our taking with us a single scrap of paper, cardboard, or anything else on which we might write a message in the hope of throwing it out of the train during our journey. That was why they ruthlessly took away our photographs. I can still see the great pile of them on the stone floor of the yard.

If, today, a film director were to show such a heap in close-up, he would certainly be accused of striving for a forced effect—especially if he were also to show a soldier's heavy boot trampling on the pile of cards, from which little girls in ribbons and boys in short pants looked up at their criminal mothers. The critics would say, "That's too much." Nevertheless, that is exactly what happened. One of the warders had to cross the yard and, rather than walk around the pile, stamped straight across the faces of our children. I saw his foot in close-up, as though it were in a film. My own were there, too: a photograph taken since my imprisonment, and the last one of them together, before they were all deported to different places. . . . When we asked if we would get the photographs back, no one answered.

"Come on, look sharp!" We were drawn up in ranks before moving off. Our matchsticks of legs dragged along the ground and slipped out of the huge regulation boots. Each of us clutched her companion feverishly for fear of losing her. The unaccustomed stay in the open air had made us dizzy and weak. Our heads were spinning, and everything seemed unreal. Fortunately, we were lightly loaded, carrying our coats only: our bundles would have been too much for us. . . . The march began.

"Not so fast in front! Close up behind!" Our old acquaintances, the Black Marias, were waiting for us. This time, however, we were not shut into separate cages but crowded higgledy-piggledy into each van, as many as it would hold. We drove through the gates. It was about sunset, just as it had been on that summer's day when we entered the prison two years before.

Through a chink in the door of our van we could see the whole of the solitary-confinement block—a close-up of our "Schlüsselburg," a three-story sepulcher of dark-red brick with high wooden shutters instead of windows. Had I really spent two years there and come out alive? In those days two years seemed to us a very long time—we had not yet learned to think in decades, nor had we heard the Kolyma saying: "The first ten years are the hardest."

We did not yet know where they were taking us, but the ominous word "Kolyma" was already in the air: it was heard in anxious questions and in reminiscences of conversations at Butyrki two years ago. At this time, however, it did not terrify us unduly. Ignorance is bliss. . . .

The freight train that awaited us at the station was just like any other, except that one of the red cars had "Special equipment" written on it in large white letters.

"Hurry up, get in! Five at a time!"

I had time to notice that my car was Number 7. It was a freight car and we were crowded in so that there was scarcely room to stand. But this cheered us up, in view of the prison rule: the more dirty, crowded, and hungry you are, and the more unpleasant the guards, the more likely you are to stay alive. So far, the rule had proved true, so thank heaven for our cattle-car conditions and the guards' rudeness. Good-by to Stolypin cars, solitary-confinement cells, punishment cells, and polite Vulturidzes!

There was a crash as the door of the car was barred with an enormous bolt; then a jerk, and the train was moving. . . .

Part two

• *Car Number* 7

Before we got into the car I had noticed the inscription "Special equipment." I thought at first, naturally enough, that this had been left on it from its previous journey. But I began to have doubts—and so did the rest of us—after the officer in charge of the convoy had told us the rules to be observed during the journey.

"The 'special equipment' must be us," said Tanya Stankovskaya, climbing onto one of the wooden bunks. "That's why they tell us that we can talk while the train's moving but not to make a sound at stations. He said we'd be put in irons if we so much as whispered."

From behind, Tanya looked like a sprightly, mischievous girl. The movements with which she folded her prison coat into a pillow were carefree and youthful. Her voice, too, sounded young as she cried from above our heads:

"Look at me! I've taken the top bunk—there's social consciousness for you. At least there's room for me here—there wouldn't be for any of you plumper ones!"

No one replied; very few of the women even heard her. Car Number 7 was full of no less than seventy-six women, all talking at once, jostling one another and thrown about by the motion of the train. All wore the same dirty-gray uniform with the brown "convict's" stripes on the jacket and skirt.

None of us stopped talking for a single moment. No one listened to anyone else, and there was no common theme: each of us talked about her own affairs from the moment the train left Yaroslavl. Some began to recite verses, sing, and tell stories even before installing themselves on the wooden bunks. It was the first time for two years that we had been surrounded by fellow human beings, and every one of us was rejoicing in the sound of her own voice. In Yaroslavl, where the prisoners came from all over the Soviet Union, those in solitary confinement had virtually not spoken for seven hundred and thirty days. For all that time they had heard some six or seven words a day: Get up, hot water, walk, washroom, dinner, lights out. . . .

In the general crush I found myself on one of the lower bunks. There was scarcely room to move, but with my experience of such matters I realized that I had got one of the better spots. In the first place, I was at the end of a row, so that I could be pushed from one side only; and secondly, I was near the high, barred window, which let through a thin stream of air. I stopped talking for a moment, raised myself on my elbows, and breathed in deeply. Yes, I could smell open fields. It was the month of June— the glorious, sweltering June of 1939.

Then, like the others, I went on talking and talking— in a hoarse voice which faltered every now and then, trying to tell everything at once and occasionally making an effort to hear and understand others. . . . My head spun painfully as I caught fragmentary sentences:

"Well, whatever happens we're lucky to be out of that stone morgue."

"Ten years' imprisonment and five years' deprivation of rights. Everyone here's got the same."

"Could you eat that soup they gave us yesterday? I couldn't, it made me sick."

"Have you heard that Anna's supposed to be here, the partisan girl, one of Chapayev's machine gunners?"

"D'you think we get anything to eat on this trip?"

Tanya Stankovskaya dangled her incredibly thin legs from the upper bunk—she had no calves, and her enormous prison boots were many sizes too big To my amazement I saw that from the front she did not look like a girl but like an old woman, with unkempt gray hair and a bony face, and her skin was dry and peeling. To my question, she replied that she was thirty-five.

"That surprises you, does it? I mean, of course, thirty-five ordinary, calendar years. If you count the two Yaroslavl ones as twenty, that makes fifty-five. Plus at least ten for the interrogation, so I'm sixty-five really. . . . Could you make room for a minute? I'm coming down for a breath of air."

She climbed down and sat on the floor next to the car door, which was open by about a hand's breadth. However, at that moment the clatter of the wheels slowed down and the guards ran up hastily, slamming the doors to and fastening the heavy wooden bar, which remained in position except when the guards had to enter the car.

"We're stopping!" At once there was dead silence, as though the carload of women had been gagged. All seventy-six of us—excited, disheveled, still afraid to believe in the change in our fortunes—became absolutely silent, doing our best to express in looks what we had been about to say. Only the more impatient ones tried to continue their conversations by means of gestures, pantomime, or even the prisoner's wall alphabet.

When the train moved on half an hour later, we found that we had all lost our voices and were speaking in hoarse whispers.

"Laryngitis—acute laryngitis," laughed Musya Lyubinskaya, who was a doctor and one of the youngest of our company. Her black braids were familiar to many of us from Butyrki.

The only person whose voice had not given out was a strapping girl from the Urals called Fisa Korkodinova. She had a metallic contralto voice with occasional bass notes, which now boomed out like a trumpet amid the confused squeaks of an amateur orchestra. This was one reason that prompted us to choose her as our starosta *— this, together with her composed manner, her rich Ural accent, and the ruddy complexion which she had not lost even at Yaroslavl. Presently we each received from her generous hands an earthenware mug without a handle, a tin bowl, and a chipped wooden spoon.

"Do you mean to say we can't smoke? Yaroslavl was bad enough, but at least they let us do that sometimes." This was Nadya Korolyova, a forty-year-old woman from Leningrad, who was almost as emaciated as Tanya but more trim in appearance, with smoothly combed hair.

Everyone began to explain that it was because of the paper. We might unroll it and write messages, and "they" were terrified of our doing so and throwing them out of the train.

I had acquired the habit of smoking in prison, and I was secretly glad that it was not allowed now—otherwise how should we have been able to breathe?

The officer in charge made his appearance, and we were relieved to see that he was a different type from the Yaroslavl warders, who spoke six words a day and stalked about the carpeted corridors like tigers. This was a cheerful rogue whom we nicknamed the Brigand, with a lock of

---

* Starosta, literally "elder"; originally, spokesman in the Russian village commune with whom the authorities dealt. Now applied to a representative of any group.

hair brushed back stylishly and a fund of racy expressions.

"Starosta of Number 7—stand up and be counted!" he roared, scanning the benches with piercing eyes and grunting with pleasure when the plump Fisa rose before him and boomed in her Ural accent: "Starosta of Number 7 reporting, Citizen Officer." Then, with relish, he repeated the various prohibitions:

"Whenever the train stops, not a squeak out of you, or you'll be put in irons. No books on the journey—you've done enough reading in the last two years. Tell each other fairy tales if you like. As to food—we-ell, you might say we're on iron rations. Back there you got two hot meals a day, here you'll get one. There's enough bread, but we're a bit short of water. You'll get a mugful a day, do what you like with it—drink, wash, brush your teeth. . . ."

"Why did you let him look at you like that?" growled a voice as the Brigand moved off. It was that of Tamara Varazashvili, "Queen Tamara," who now tossed her proud head even higher. She had been arrested in 1935; her father, an expert on Georgian literature, had been accused of nationalism. She was guilty of nothing more than being her father's daughter, but she considered herself a "genuine political" and coldly despised the "1937 lot" for their lack of independence, their wheedling tone with the guards, and the fact that they asked for things instead of demanding them.

"Like what?" asked Fisa in a puzzled tone.

"You know perfectly well—he was ogling you. And how could you smile back at him? It shows lack of self-respect."

Seventy-six hoarse voices began arguing at once, as usual paying no attention to one another. Then the voice of Polya Shvyrkova prevailed:

"After all, some of them are human beings. Why shouldn't he make eyes at her? She's a good-looking girl, and I think it's a good sign if he wants to—it means he sees us as human beings, even as women. And it's better to be any sort of woman than a number, isn't it?"

At these words the rest of us fell silent for a moment. It was as though we who had been buried alive in a dank dungeon till yesterday, who only this morning had been given back our names instead of numbers, had once again felt a cold breath from the tomb.

"Bravo, Polya! You're quite right—anything but a number!"

"I'm sorry if I've spoken out of turn—you're all clever people and Party members, and I'm only a cook. They arrested me because of my relations—I don't know why they had to think up anything as fancy as counter-revolutionary Trotskyist activity."

Somehow or other the day wore to a close. A new moon was visible through the barred window. There were two or three more bursts of conversation, and then silence.

I lay down on the bunk. It was not uncomfortable—in this heat one was glad of bare boards, and the prison coat made an excellent pillow.

"Tell you one thing," said Tanya's voice from above, "if I were a queen, I'd make sure I got the bottom bunk every time."

My neighbor introduced herself to me: Zinaida Tulub, a Ukrainian writer of historical novels. When she asked who I was and what I did, I could not reply at once. Until today I had been "Cell 3, north side." Finally I told her my name and said I had been a teacher and journalist. As I heard my own voice I felt bewildered, as though I were speaking of someone else. Could it really be me? A girl at Butyrki called Sonya used to reply to questions about her past: "It was long ago and it never happened anyway."

I was almost asleep, oblivious at last to the excitements of this incredible day, when I felt something furry brush my face. What was this? Was I back in the punishment cell with the rats? Was Car Number 7 boldly marked "Special equipment" just a dream after all?

"I'm sorry, comrade, my hair brushed against you."

Zinaida Tulub looked like a great lady of the last century. Her hair was magnificent, bedraggled and dirty though it was.

"Did I frighten you, comrade? You're crying!"

No, I was not crying, but my heart was pounding with excitement. I so wanted her to go on and on calling me "comrade." To think that such a word still existed and that someone could use it to me! So I was not just Cell Number 3, north side, after all. The train was bound eastward, toward the camps. "Penal servitude—what bliss!"

2

• *All sorts to make a world*

Or, as the Germans say, God has a big menagerie. I thought of this constantly as I became acquainted with my companions and fellow travelers in Car Number 7. How fantastically varied they were!

Our day began early—the Yaroslavl habits were so ingrained that our tiredness made no difference. When Tanya sat up in her top bunk, bumped her unkempt gray head against the ceiling, and barked "Everybody up!" we did not need to be told twice.

Our excitement of yesterday had died down a little; we felt more down-to-earth as we looked about us. Some of us recognized people we had met at Butyrki, or even in the outside world. Fisa, our starosta, had doled out to us our first ration of bread, with extra pieces neatly skewered to it by little sticks of wood. The day's routine was visibly taking shape.

On the first day we had been so impressed by the fact of moving at all that we did not notice how slowly the train was crawling. Now we saw that it was going at the pace of a slow-motion film, or the kind of sledges in which the Decembrist wives drove to rejoin their menfolk.

The train creaked agonizingly; the wheels rattled unevenly, and kept splashing the precious water in our mugs. We were forever stopping at stations and lesser halts and even in open country so that the guards could bring food along or carry out inspections. Joyfully we set about applying the techniques of cell life we had learned at Butyrki, such as watching by turns out of the little barred window. Many gave up their turn at this to Anya Shilova, an agricultural expert from the Voronezh region, knowing how much she longed to look at crops growing in the fields again. Anya was a short, stocky, somewhat ungainly figure, a sound sleeper and bursting with energy. She looked out of the window and said with a patronizing smile:

"Yes, of course, they are not up to much in these parts. You should just see the crops around Voronezh at this time of the year."

And Tanya Krupenik, a Ukrainian to her fingertips with her brown eyes and black eyebrows, exclaimed:

"Just think, Anya dear! We'll soon be working, and how lucky you and I are to be agronomists! Perhaps Musya will get a chance to work at her own job too, but it'll be harder for those with a non-technical education."

"How do you know what'll happen to us? None of us has ever been in a camp." This was Sofya Andreyevna Lotte, a specialist in West European history from Leningrad and the only one of us seventy-six women who was not wearing the "Yezhov uniform." She wore a Leningrad knitted blouse and black skirt, shabby but her own. This fact made some of the others look at her suspiciously, though it was not clear what harm she could have done anyone in conditions of solitary confinement.

"No, of course," put in Nina Gviniashvili, the Georgian actress. "None of us has ever been in one, and it's quite possible that you, dear Lotte, will find a professor's chair waiting for you. Then you can teach the polar bears about the Chartist movement."

Nina was as much given to this sort of leg-pulling as Tanya; but while Tanya's comments were generally on the bitter side, Nina's were lighthearted and in good taste, in the French manner, so that not even the victim was offended. This may also have been because Tanya was thin and untidy, with sharp cheekbones and a peeling skin, whereas Nina was an elegant woman of character who was growing old gracefully. Perhaps her most striking features were her eyes, which shone a bright green in the gloom of the car.

In actual fact, most of the women in our car had an arts rather than a science background, and they loved to listen to Zinaida Tulub reciting her own poetry. This she did in an old-fashioned, rhetorical way, with an almost ecstatic expression. Her whole manner of reciting, and even talking, indeed suggested the literary *salon* of pre-revolutionary days. In practical matters she was helpless and often cut a rather comic figure. According to Tanya, she had two loved ones "outside": her husband and her cat named Lyric. She had asked the governor at Yaroslavl to transfer fifty rubles from her account for the benefit of Lyric, who could not live without meat.

Zinaida had been better off in solitary confinement, where she could daydream and compose poems. She had not once been put in a punishment cell—could this have been out of respect for literature? . . . Here, on the other hand, she was apt to be pushed around. She was older than most of us, our average age being about thirty.

But poetry is a common bond for everyone. At Yaroslavl I often used to fancy that I was the only one who sought and found in it a way of escape from the closed circle of our lives; that I was the only one to be visited by Blok in my cell, the only one who repeated to herself during her solitary walk, "I want one poison only—the poison of poetry." But this was presumptuous of me, as I now realized, listening to the flood of verses that we all poured out: clever or simple, lyrical or denunciatory.

Anya Shilova, leaving the harvest to take care of itself for a while, sat curled up on the middle bunk and declaimed to her companions Lermontov's poem about the hapless Demon, "spirit of exile." Never before in her life had she learned such a long poem by heart. As she recited it, her voice thrilled not only with expression but with the pride she felt at being in contact with such works.

Polya Shvyrkova, the cook, had pored over Nekrasov and had begun to make up verse of her own. She was, as a matter of fact, rather flattered at being charged with such an "intellectual" crime as Trotskyism and, as a result, brought into the society of Party members and scholars. Even at Butyrki she had picked up a number of impressive new words: the expression "state of affairs" was one of her favorites. She would say: "Well, Genia, what do you think of the state of affairs now that Yezhov's gone? Things are looking up, aren't they?"

Suddenly the poetry lovers were joined by our former "special correspondent," Olga Orlovskaya, and I was dumfounded to hear her recite the fulsome lines in praise of Stalin which I have already quoted. She had, she informed us, shown them not long ago to Vulturidze. . . .

"Well," came Tanya's rasping voice, "did he burst into tears and fall on your neck?"

This set off a terrible argument. In spite of all the evidence, at least twenty members of our company asserted with lunatic insistence that Stalin knew nothing of the illegalities that were going on.

"It's those hellhounds of the secret police," said Nadya Korolyova, "and Yezhov, whom he trusted so much. But now Beria will put things right, you'll see. He'll prove to Stalin that we're all innocent, and we'll soon be let out, you mark my words. We must all of us write to Stalin so that he knows the truth, and when he does, how can he let such things happen to the people? Take my case, for instance . . ."

She was interrupted by Chava Malyar, a handsome woman of about forty, who looked like Verdi's Aïda; she had joined the Communist Party before 1917 and had once been sentenced to death by White Guards.

"It won't do, Nadya dear," she said with a smile. "You're a true daughter of the Petersburg working class that used to believe in the wicked ministers' fooling the good kind Tsar; but they stopped believing it on the ninth of January

1905,* and now it's 1939. But you're talking like one of those Zubatov people—that won't do nowadays."

Genia Kachuriner and Lena Kruchinina now leapt into the fray to give a "scientific" explanation of everything going on in the country. The stifling air in Car Number 7, with its piles of prison coats and earthenware mugs still containing some muddy-looking lukewarm water, vibrated to the sound of arguments about the intensification of class warfare in the process of building socialism, objective and subjective collusion with the enemy—and, of course, the notion that you can't make an omelette without breaking eggs.

"That's enough from you bookworms," shouted Tanya Stankovskaya furiously from her top bunk. "What do a couple of Moscow young ladies like you know about the working class?"

Lena replied sharply: "A strange remark from someone who claims to have worked in the Party apparatus."

"What do you know about the Party apparatus, anyway? Have you seen ours in the Donbas, for instance? Have you ever seen anything outside Moscow? Cut it out, for Christ's sake. Let's sing something instead—at least we're allowed to do that here." And Tanya, in a mock-drunken voice, struck up a parody of a well-known marching song:

The convicts marched
In fear and trembling
All the way to Siberia . . .

* "Bloody Sunday," on which a peaceful crowd led by the priest Gapon, attempting to present a petition to the Tsar, was fired on by the police in St. Petersburg. Zubatov was the Moscow chief of police early in the century, who experimented with "Police Socialism." "Zubatov people" refers to those workers who believed that this was a genuine trade-union movement to improve the lot of the people.

For Trotskyism, terrorism
And talking politics.
But God alone knows
What they'd really done wrong!

"We ought to sing something more serious," said the
Ukrainian Tanya, "about our men, and whether they're
alive or not."

And so with parched throats we sang:

Husbands, lovers, and friends,
Listen to the song we sing.
We your women are following you,
Swiftly as birds we fly.

Our husbands . . . Some of us knew for certain that
they had been shot. Others knew nothing, but all of these
cherished hopes of a future meeting, and the song was
followed by eager whispers.

"I'm sure we'll come across them. Everybody is put together in the camps."

"Perhaps they'll even allow us to live together. Lucia
was saying . . ."

This was the point where the Menshevik and Social
Revolutionary women, who had had long experience of
prison, came into their own; they knew what was what.
Lucia Oganjanian, a shortish woman much like Rosa Lux-
emburg in appearance, with a long nose and clever, mocking eyes, liked to talk about the best time of her life, which
she had spent in the early thirties at the political prison
of Verkhneuralsk, and where "liberalism" had gone so far
as to permit the sharing of cells by husbands and wives. She
blushed and her face became prettier as she told us: "You
only know about pure, disinterested love from literature,

but I have known it in real life. I can't complain, I have had my share of happiness. That magical solitude, a real coming together of two souls that is only enhanced by the hostility of the surrounding world . . ."

Katya Orlitskaya was a Social Revolutionary, a woman of over forty, with straight strands of hair and an almost masculine-looking, aquiline nose, like an Indian chief. She assured us that we would certainly meet our husbands at the transit camp. We did not, of course, yet know where they were taking us; but if it was to Vladivostok, then we were sure to meet them there.

"Don't listen to them," whispered Nadya Korolyova. "What do those old dears know about it all? They and their so-called socialism! When we learned about them at our political study groups, I thought they'd died out a long time ago and now they turn up here. Some company! All I can say is, damn the investigators who lumped us together!"

The song about our husbands tempted us to talk about the forbidden subject of our children. Time and again, even at Butyrki, we had agreed that we must not; but there was always someone who could not help herself.

"I remember when he was four years old, I bought a chicken and killed it, and he said, 'I can't eat that one, it's a friend of ours.'"

Zoya Maznina, the wife of a Leningrad Komsomol leader, had three children, of whom she most often mentioned the youngest, Dimka. She smiled instead of crying when she spoke of him, and remembered one funny incident after another.

The floodgates were open, and we all gave way to our recollections. The twilight gloom of Car Number 7 was filled with children's smiles and tears and their voices saying, "Mother, where are you?"

Some lucky ones had had letters at Yaroslavl saying that their children were alive and with their grandparents. Others, like Zoya herself, had heard nothing for more than two years. Her husband and brother had been arrested, there were no grandparents, and she had no one to write to. It was doubtful if strangers would take the risk of writing to her, and in any case the Yaroslavl rules forbade correspondence with anyone but close relatives, and even so it was very restricted.

I was one of the luckier ones, since I knew from my mother's letters that one of my children was in Leningrad and the other in Kazan, with relatives who I hoped would treat them kindly. My mother had copied out some lines from a letter of Alyosha's: "Dear Granny, The mother of the boy who shares a desk with me is in the same box as Mother." (In Yaroslavl, our only address was a P.O. box number.)

I clenched my fingers to keep from breaking down. But there was no stopping the outburst that was about to engulf Car Number 7. I remembered the same thing from Butyrki. Once it began there was no stopping it. Soon we were all sobbing and crying: "My little boy! My darling little girl!" . . . After such outbursts, we thought obsessively of death. Better a horrible end than to go on suffering like this forever.

No, it was no use letting oneself go. At least I was a thousand times better off than Zoya—or Milda Kruminsh, who in two years at Yaroslavl had got only one letter from her Yan. He had written, from one of the homes for prisoners' children: "Dear Mother, I am having a good time. We gave a concert on the First of May." And then, hastily scrawled underneath: "Mother darling, I've forgotten the Latvian alphabet, please write it out for me and I'll write to you in Latvian, I don't like it here without you."

If only we could learn, like Zoya, to smile instead of crying when we talk about them! . . . I told them a little rhyme Vasya had made up when he was three years old. Another time, instead of an outburst of grief there was a row because Zinaida Tulub had butted in with some reminiscences about her cat Lyric. She was attacked, almost bodily, by that same Lena Kruchinina who had given a "scientific" explanation of the events of 1937. "How dare you?" Lena shouted in a very different voice from her usual academic one. "How dare you put on airs like that? There are mothers here, do you understand?—Mothers whose children have been snatched from them and left to the mercy of fate. It's an outrage to compare our children with your cats and poodles!"

Luckily, there was a stop just then: the uproar died down instantaneously, the "special equipment" held its breath. Only Nina Gviniashvili dared to remark in a barely audible whisper:

"Very well put, Lena! Very precisely formulated! But how does the problem of children tie in with the intensification of class warfare in the process of our advance to socialism?"

Through our little window, we became aware of the smells and sounds of the summer evening. At Yaroslavl it would have been nearly time for lights out. It was clear we should not be getting any more water, though we had hoped for some till the last minute. Even Fisa Korkodinova, our starosta, said: "I thought the Brigand was only trying to frighten us when he said a mugful a day. It's only about half of an ordinary glass. What are we going to wash with?"

Tamara Varazashvili tossed her head as usual and said in a low voice:

"They would never dare torture us with thirst if we'd demanded water instead of humbly asking for it. Why should we be polite to these people?"

This started another argument. Sara Krieger, whose throat was so sore she could hardly speak, explained patiently that it might have been all very well demanding things in a Tsarist prison, for instance, but in one of ours . . .

"Oh, really, so you have your own prisons, do you?" put in Nina Gviniashvili coyly.

At this point Nadya Korolyova—the gentle, obliging, easygoing Nadya—decided she could stand no more. She cried out:

"Oh, do keep quiet and let's get some sleep. The trouble with you intellectuals is, you quarrel so much. All right, so there's not enough water—it can't be helped, we must put up with it. The main thing is, we're going to work, and not staying in that great stone morgue of a place. . . ."

(Four years later Nadya Korolyova, returning from the day's work in the violet evening light of Kolyma, collapsed on the icy ground and by this held up the rest of the work gang. The four others who marched in the same rank cursed her, and the guard, prodding her lifeless body, shouted: "Stop playing the fool! Get up, I tell you!" This went on for some time, until one of the prisoners said: "But she's . . . Can't you see?")

Nearly all of us were now asleep. I was in acute discomfort, afraid to turn in case I should wake up Zinaida. The poor thing had wept so helplessly when Lena had snapped at her like that. . . .

There were some people awake on the middle bunks too, where the women from the Caucasus were talking softly to two Germans from the Comintern. Tamara was delighted to be asked questions about Georgia by Maria Zacher, a journalist and member of the German Communist Party. Dreamily, to the sound of the wheels, Tamara described her country.

"We have an ancient culture. The country was Christianized in the fifth century. . . . Shota Rustaveli,* the poet . . . A proud, fearless people fond of good living . . ."

"Not to say bone-idle," put in Nina Gviniashvili.

This brought a titter from Lucia Petrosyan, the Armenian, who was the sister of the legendary Kamo.† She used, however, to avoid mentioning this, and when Nina urged her in ringing tones not to conceal her brother's name but to be proud of him, she would reply with false humility: "I'm only a plain, ignorant girl from the mountains." The reason for her caution was that she had once been personally acquainted with Stalin, and now lived in the hope that, one morning or the next, the officer in charge would open the door and summon her, "with her things," to freedom. She was now in her third year.

Tamara was cross with Nina for spoiling her romantic picture of Georgia. "You're nothing but a renegade—and yet with those eyes, you can't deny you're a Georgian. Look at the way they shine!"

(This is what was to happen to the splendid green eyes of the clever and elegant Nina Gviniashvili, who had been an actress. At the Kolyma collective farm of Elgen, where some thick branches got into the silage cutter, the cutter

---

* Twelfth-century Georgian poet.

† Pseudonym of Simon Ter-Petrosyan, famous for his daring part in the bank robberies ("expropriations") used to finance the Bolshevik Party before the revolution.

got out of control, as it was in the habit of doing, and a
tough, thorny piece whipped out Nina's right eye. . . .
When Pava Samoylova and I went to the camp infirmary
in order to take her some sugar, and sat at her bedside, so
upset we could say nothing, she stroked Pava's hand and
said: "Don't worry, girls—one eye is quite enough for
looking at a life like ours.")

It was far into the night before I could sleep, even after
Maria Zacher had rounded off the conversation about
Georgia with a few earnest Russian sentences constructed
in the German way with the verb at the end:

"Thank you, Tamara. These are facts that for my
knowledge of the Soviet Union of great importance will
be."

Next morning we all felt embarrassed about the arguments of the day before.

"Listen, girls! Don't you remember how we longed,
back in Yaroslavl, to know who our neighbors were and
to talk to them? And now we do nothing but quarrel.
Why should we take our troubles out on each other like
this?"

We were all touched by these words, especially as they
came from Pava Samoylova—or Pavochka, as we always
called her. She had close-cropped hair and clear round
brown eyes, and she had been caught up in the witches'
Sabbath because of her brother Vanya, who had once
belonged to the opposition in the Communist youth movement. Her life had two chapters: school and prison. She
looked rather like one of those girls who joined the revolutionaries in Tsarist times.

"Wouldn't it be a better way of spending the time if
each one told us about her own special subject? Anya about

farming, Musya about medicine, Sofya about history. And you, Genia—you can recite Pushkin, you know him by heart!"

"That's a good idea. Let's have some classical poetry—it's the most soothing thing there is!"

It appeared that I was very good at this even by Yaroslavl standards, and also that there were material advantages in reciting poetry. For instance, after each act of Griboyedov's *Misfortune of Being Clever* I was given a drink of water out of someone else's mug as a reward for "services to the community." My own mug remained covered by a small dish, and I looked forward to the evening, when there would be at least a quarter of a glassful left in it.

Then we came to Nekrasov's "Russian Women." How often, at Yaroslavl, our thoughts had turned to the Decembrists and their wives! I recited the passage about Princess Volkonskaya meeting her husband:

> I fell on my knees to him. Lifting his chains,
> I kissed them before I embraced him.

No, these were no longer just lines in an anthology—they expressed the longing of all seventy-six of us. As I declaimed them, I saw before me all those pairs of anguished eyes. . . . As for the Decembrists' wives, we felt as if we were sharing the journey with them and they with us. None of us would have been surprised to find Marya Volkonskaya and Katya Trubetskaya sitting next to Pava Samoylova and Nadya Korolyova. But of course, they had had an easier time of it:

> Of wondrous build, so light, so strong,
> The little wagon speeds along.

That was more than we could say of Car Number 7! But here was Tanya Stankovskaya sighing from her upper bunk: "Never mind, girls, I'd walk to Kolyma like a shot if I knew that my Kolya was there!"

Yes, Marya Volkonskaya had been very lucky when she met her Sergey in the mine:

> Oh, sacred that silence, how sacred! 'Twas filled
> With some sort of grief past betraying,
> Some deep, solemn thought . . .

I went on reciting line after line. After a while I was aware of dead silence, and my voice sounded strangely loud. Suddenly I realized why—the wheels were no longer accompanying me. The train had stopped!

Good God, what have we done? The rule about absolute silence at stops . . . What will happen to us? How do they punish "special equipment" for talking?

The door crashed open, the Brigand's voice rasped out: "Give up that book at once!"

This time he was not joking, or looking for Fisa Korkodinova: he looked more like one of the Yaroslavl guards, the Nabob even. . . .

"Give up that book, I tell you! Starosta of Number 7, what are you up to? If you don't give up that book at once, by God I'll give you a frisking that you'll never forget. Mishchenko, find that book, or I'll have them all in irons. I'll teach them to break the rules and make a circus out of the convoy. . . ."

"Move up to one end of the car, all of you!" ordered the flat-nosed Mishchenko, pushing us all to one side and rummaging with quick, professional movements through the brown-striped Yaroslavl coats.

Tamara, whom he had given a particularly vicious push, said in a quiet but angry voice:

"I protest. No one has broken the rules. There are no books in this car. Our comrade here was reciting from memory."

This roused the Brigand to a frenzy.

"What the hell do you mean by trying to make a fool of me? No book? But I stood beside the car myself for a whole hour and heard you reading out of it."

"No, it was from memory."

"Well, of all the damned gall! Listen, just for telling such lies you'll travel the rest of the way in irons. I'll teach you to make fun of the commanding officer! I tell you for the last time, give up the book, or you'll only have yourselves to thank for what'll happen to you!"

Fisa, our level-headed, quick-witted starosta, saved the situation. Standing rigidly at attention and keeping her voice with its Ural accent under control, she said: "Excuse me, Citizen Officer, may I suggest that you hear for yourself? If you make her recite, you'll see what a lot she knows by heart. She has a remarkable memory, we could hardly believe it ourselves. She could go on the stage with it. Do please try it, Citizen Officer."

The Brigand struggled visibly with his feelings. He was afraid of being made a fool of, but the reference to the stage had struck home. Reading without a book! Well, by God, you never knew with these intellectuals. Perhaps she really could. . . . Finally curiosity got the better of him.

"All right," he said. "Come on, Mishchenko, we'll find out if it's true. Which one is it—the dark one over there? All right, fire away. I'm going to time you by my watch, and if you can go on for half an hour without a break,

O.K.—otherwise it's irons for you, all the way to Vladivostok!"

The wagon hummed with joy and relief. In the first place, the Brigand in his excitement had blurted out our destination. From Vladivostok, no doubt, we should go on to Kolyma, with its opportunities for heroic work and early release. Secondly, everyone knew that the "stage turn" would be a great success, as they had already seen it done.

"All right, start!" ordered the Brigand.

"Won't you please sit down first, Citizen Officer," said Fisa in her homely manner. "Otherwise you'll get tired before it's over."

"Very well. Come on, Mishchenko. Now, let's hear."

"The Russian Women"—no, obviously I wouldn't recite that to them. Something neutral. "Eugene Onegin. A novel in verse, by Alexander Sergeyevich Pushkin."

As I went on reciting, I kept my eyes fixed on the two guards. The Brigand at first wore a threatening expression: she'd get stuck in a minute, and then he'd show her! This gave place by degrees to astonishment, almost friendly curiosity, and finally ill-concealed delight.

"Well, I'm damned! You Trotsky-Bukharinists are a clever lot, there's no doubt about it! Think of that, Mishchenko—every single word by heart. Hey, what are you stopping for? Is that all you know? No? Well, go on, then!"

So I went on. The train had started again, and the wheels kept time to Pushkin's meter. Someone offered me precious refreshment—an earthenware mug partly filled with warm, dirty-looking water. "Do have some of mine, or you'll lose your voice."

The flat-nosed Mishchenko, lulled by the rhythm, had

gone to sleep: every now and then he started and shook himself. As for the Brigand, with his ready curiosity and love of mischief, he was an excellent listener, laughing and being moved in all the right places. He was delighted with the description of the Larins' guests:

> Those gray Skotinins, who surprise us
> With offspring of all ages, sizes,
> From two to thirty at the top.

"Great stuff!," he leered, interrupting me and squinting at Fisa. "From two to thirty, eh? They didn't waste much time!"

There was no more talk of putting us in irons. When I had finished, the Brigand said:

"When we get to Sverdlovsk you can have a bath. It's the regular disinfection point. All the water you want for washing and drinking. It won't be long now."

"It won't be long." We heard these words every morning, but they proved deceptive.

"I honestly believe we must still be somewhere in the Yaroslavl district," said Anya Shilova in despair, as she came away from the window. "Here, Mina, you're a geographer, you have a look."

Mina hoisted herself with difficulty onto the middle bunk. She was about fifty, and was a Party member of long standing. The years of imprisonment had made her shrunken, bent, and as yellow as a lemon. She complained a good deal about her ailments, and this was not popular in the car, where the unwritten law was "Suffer in silence." We all of us in fact had plenty to complain of. Tanya Stankovskaya, for instance, was a walking skeleton and ghastly to look at, but she never said a word about her troubles.

So no one paid attention when Mina Malskaya, after clambering onto the bunk, clutched at her heart. Instead, we asked: "Well, where does it look as if we are going?"

"Well, it's hard to say. If the plant life is anything to go by . . . but in this heat . . ."

Compared with Mina, Anya looked young, strong, and full of energy. She was thrilled by the prospect of work. Provided she was allowed to do something in her own line, she didn't mind where the devil they were going. "You know, girls," she would say, "I wouldn't mind digging the frozen ground with my bare hands—after all, it's our native soil. The one thing I can't bear is to sit idle. If they'd kept us another year in solitary, I'd have bashed my head against the wall. I can stand anything except idleness."

(In 1944 the lives of these two women, both loyal members of the Party but in other respects so unlike each other, ended in almost identical ways. Anya Shilova died in a camp infirmary of a kidney disease brought on by overwork, and before her death she experienced the horrors of blindness. Mina Malskaya died a month later of a heart attack. Three days afterward, a telegram came for the camp commander: "Please provide assistance necessary to save my mother's life. Boris Malsky, *Izvestia* war correspondent.")

Anyway, we knew we were going to Vladivostok, and the Menshevik and Social Revolutionary women assured us there was a transit camp there from which we would most likely be sent to Kolyma. But that was a long way ahead, we needn't think of it just yet. If only we could see the last of Car Number 7, and have a decent drink of water!

The excitement of our first days together had subsided: we no longer argued about abstract topics or even recited poetry. We all realized the grim truth: that at this rate it would take more than a month to get to Vladivostok.

It was hot and stuffy—so stuffy that one of us hit on the expression *dushegubka*, which was later applied to the German mobile gas chambers. We suffered from dust, sweat, and lack of air, but most of all from thirst.

"Girls!" said Pavochka once with a look of surprise and pain on her childlike face. "Why do they say that no one is a hero who cannot conquer sleep? I think it should be 'No one is a martyr who has not been tortured by thirst.'"

Hardly anyone ate the salted soup which was doled out to us and in which herring tails had been boiled. Although liquid, it made us frightfully thirsty, and almost all of it was handed back to the guards untouched.

One day Fisa appealed to the commander on behalf of us all. "The water that's used for the soup—could we please have it simply by itself? Then we could at least wash the dust out of our eyes. A mugful of water a day's nothing. We are really on our last legs. They must use four or five gallons for the soup, and then nobody wants it. With that much we could wash ourselves and our clothes and have some to drink as well. After all, Citizen Officer, we're women. . . ."

This put him into a rage.

"Let me tell you, starosta of Car 7, that neither you nor I are entitled to change the diet as we see fit. The orders say hot food once a day, and that's what you're getting. If I gave you water I know what would happen, you'd complain to Gulag * afterward that we'd done you out of your nice hot soup and eaten it ourselves. Especially

* Chief camp administration. The department of the Secret Police (NKVD) which was responsible for all concentration camps in the Soviet Union.

as you're such a damned educated lot, so good at writing.
No, you'll get the regulation diet and that's that."

Tanya Stankovskaya was a terrible sight. Her skin was
peeling more and more; her teeth had grown long and
irregular and stuck out between her chapped lips like the
palings of an old, tumble-down fence. And although she
still never mentioned her health, it was clear that she had
frightful diarrhea. Twenty times a day she would climb
down from the top bunk and make her way—gray, disheveled, and ghastly to look at—toward the corner of the car
where a large hole gaped in place of the prison slop bucket.

"Musya—you're a doctor, can't you do something for
her?" Musya shrugged her shoulders and tossed her head.
"Oh, you liberal arts people! What kind of magician do
you think I am? All right, supposing I prescribe a vitamin
diet, intravenous injections of glucose and antiscorbutics,
confinement to bed and, of course, plenty to drink—what
then? . . . Listen, Genia, what she's got is called pellagra,
and it means three D's. . . ." She moved her lips, trying
to remember. "My God, I'll forget everything in this
place. . . . Three D's. The first is dermatitis, that's the
way her skin's peeling; the second's diarrhea, and you can't
do anything about that without vitamins."

"And the third?"

"The third—well, I don't think she's got that yet, it's
dementia. Madness—mental derangement."

No, Tanya certainly did not have that yet. I knew,
because in the evenings she used to invite me up to her
top bunk and express thoughts that were far from insane.

"You know, Genia, I can't stand hypocrites. You like
Olga Orlovskaya, but I think she's the limit. That dreadful poem of hers about Stalin—well, really! I daresay the
people who aren't in prison can still be fooled, that's one

thing, but here in Car Number 7, in the seventh circle of Dante's Inferno, anyone who goes on raving about our father, leader, and creator must be either a hypocrite or a congenital idiot."

"Listen, Tanya dear. Suppose we tell the commander you're very ill—perhaps there might be some sort of hospital ward on the train?"

"Don't try to be funny, Genia."

"No, I mean it. At Yaroslavl there was a doctor, and a very kind one too. I used to call him Pickwick. Do you remember, last year when they locked the outer windows?"

"Yes, I remember vaguely. That was when I got this way, from oxygen starvation."

"Well, I was very ill then too, with heart trouble, and Pickwick was kind to me: he arranged for us to get twenty minutes' fresh air a day instead of ten. Why don't we ask about a hospital ward, Tanya? There may be one."

"What they have got is a punishment cell in the last car —Fisa told me. As for a hospital ward, it's your girlish imagination. You and Pavochka are a trusting pair, all right —it's rather funny, you've both come out of Yaroslavl with the mentality of grown-up children. Listen, that's enough about the things of the flesh, it's past helping. I want to tell you about a secret thought of mine. I know you'll understand. . . . I can't bear to look at Nadya Korolyova. It's on my conscience. I forget that I'm a prisoner here myself."

I understood without further explanation. I had felt the same sense of shame and personal responsibility in 1937 at Butyrki, when I shared a cell with the Communists from abroad. . . . I stroked Tanya's bony, almost fleshless shoulder and whispered:

"Yes, Tanya, I understand. I felt desperately ashamed at Butyrki when I met Klara, the German Communist. She had escaped from the Gestapo by some miracle. I felt I was answerable for her being in Butyrki."

The July heat got worse and worse, and the air was so hot and humid you could feel it on your skin. The roof of the car was red-hot and the nights were not long enough to cool it. While the train was in motion, the car door was left open a hand's breadth, but whenever we slowed down before a stop the guards came along (walking twice as fast as the train was going) to shut the doors and fix the bars in place. Then, when we had left the station, the train would stop in the open country and the guards would unbar the doors and leave them ajar as before. The right to sit by the opening was governed by a strict rota. Those who, for the time being, were not entitled to sit by the door or the window lay helplessly on the bunks and moved their parched lips as little as possible.

The thoughts of each one of us were anxiously centered on our ungainly little earthenware mugs. How could one prevent the water in them from being spilled by the jolting of the car or by a careless movement of a person next to one?

Some preferred to drink their whole ration in the morning. Those who kept it to sip from time to time until evening did not enjoy a moment's peace: they were constantly looking at their mugs and trembling for them. Quarrels kept breaking out and threatening to produce a complete breach between people who had been friends the day before.

Our two theoreticians, Sara Krieger and Lena Kruchinina, lay back to back and were not on speaking terms. Sara had put her hand into her jacket pocket in order to

take out a bandage which she had got from the nurse at
Yaroslavl for her feet. She had been issued with boots far
too large for her and they had rubbed her feet raw. As
she did so she jogged Lena's mug and spilled quite a lot
of water—about a tablespoonful. Lena was so furious that
she almost struck Sara, who looked horrified and exclaimed
in her guttural voice:

"I'll make it up to you tomorrow, I really will! Please
don't make a scene. Remember, there are people here who
aren't Party members!"

But Lena could not contain her rage, and her cheeks
went bright red as she retorted: "Yes, and there are Mensheviks and Social Revolutionaries too, but that doesn't
mean you have to be as clumsy as a hippopotamus and
start spilling people's water."

(Later on, Lena Kruchinina was to do rather well for
herself in the camp. She managed to win the heart of the
commandant of the women's section, a ferocious female
called Zimmerman, who appointed her to one of the posts
of authority which were usually reserved for non-political
offenders, and never given to "enemies of the people."
Lena was empowered, for instance, to decide who would
get the cushy jobs in the camp compound. She also made
friends with the camp starosta, a criminal who went by
the name of Goldilocks. During roll call the two of them
stood alongside the commanding officer and decided who
was to do what. They possessed elegant quilted coats, felt
boots of the best quality, and warm mittens knitted by
old women in the hut for Germans. In less than a year's
time, many of our present company would be calling Lena
a "damned stool pigeon," or, as the intellectuals put it,
"the court favorite.")

Once an extraordinary thing happened: the guards

omitted to bar our door at one of the stops. Humanly speaking, this was understandable: they had their hands full all day, if only with counting and recounting their charges. This was done twice every twenty-four hours, though it was not clear how any of the "special equipment" could have escaped from the sealed cars. Anyway, this time a miracle took place, and through the partly open door we began to hear sounds of ordinary human life: laughter, children's voices, and the gurgling of water.

It was almost more than we could bear. At least twenty of us jostled and crowded close to the narrow opening. We were at a small station, exactly like hundreds of others, somewhere in the depths of the Urals. Barefooted urchins were selling eggs which they displayed in their caps. On a rusty sign affixed to a wooden shed were the words "Hot water."

For once I firmly defended my "place in the sun," and secured a good vantage point on the floor, next to the opening. With all my heart and soul I threw myself into the life of the little station, repeating earnestly to myself:

"O Lord, please grant a miracle. Please turn me into the poorest and ugliest of these old women, squatting on the platform and waiting with their buckets and pots to sell food and drink to passengers. I would never once complain about fate until the day I died. Or let me be like the hag over there, poking a stick between the dirty loose planks. . . ." Perhaps she had only weeks or days to live, but at least she was a human being and not "special equipment."

The worst torture was to see and hear a tap with water dripping from it. A youth, naked to the waist, bent over and allowed it to trickle over his sunburnt back. This was too much for someone in our car. She poked her earthenware mug through the opening in the door and called out: "Water!"

After the whole scene was over, many of us said it reminded us of an incident in Tolstoy's *Resurrection*.

"Saints above, I do believe it's a convict train," said one of the women crouching beside their pails full of cucumbers.

"Where? Where?"

"We must give something to the poor souls. Here, Dasha!"

"Bring over some eggs!"

"They're asking for something to drink. Manka, bring some milk!"

Horny, weather-beaten hands were thrust into our car with pickled gherkins, curds, eggs, and bread. Beneath the kerchiefs which covered the women's foreheads we saw the ancient peasant eyes filled with tears and pity. One of them splashed milk into the mugs we were holding out, and as it spilled it made little pools on the bare earth.

"Look, there's nothing but women."

"Perhaps there are men in some of the others—how can we tell?"

"Lord save us, do you think the Gavrilovs' lad could be there?"

"Fancy the cruel devils not giving them any water! Here, Anka, fill up the pail again."

"It's too big to go through."

"And think of all their poor little children left at home, no better off than orphans."

I felt for a moment as if we were not in 1939 but back in 1909. But the modern age reasserted itself in the person of a young woman who hastily thrust into the car a bunch of spring onions, saying: "Here, eat some vitamins. That's the most important thing of all!"

All this went on for several minutes, and by some amazing chance the guards, who were replenishing the train's water supply, noticed nothing. We moved off again. A specially elected committee consisting of Fisa, Pava, and Zoya set about dividing the onions into precisely equal portions. But even the distribution of these did not lessen our growing indignation and rebelliousness over the water supply. Before long Tamara Varazashvili, standing in the middle of the car, addressed us—without raising her voice, but in a formal tone—as follows:

"Comrades, I wish to say a few words. We must demand a sufficient ration of water. We are at the end of our tether, and all of us have spent two or three years in prison —and what prisons! We are all suffering from scurvy, pellagra, and malnutrition. What right have these men to torture us with thirst as well?"

"Quite right, Tamara!" said Chava Malyar, raising her voice for the first time since the beginning of the journey.

"Do not speak for all of us," said someone from the top bunk.

"Naturally," Tamara went on, "I do not include those who are willing not only to knuckle under, but to justify it all in the bargain."

"And to produce theoretical arguments why it must be so," said Chava, getting up and standing beside Tamara to emphasize her support.

"Moreover," Tamara went on, "what difficulty can there possibly be? We aren't traveling through the Sahara. What is to stop them from taking on water at the stations three times a day?"

"Well, what are you suggesting? A hunger strike?" came a voice from the corner where the Social Revolutionaries were sitting.

"I demand"—this was Lena Kruchinina—"that you stop

this anti-Soviet propaganda, and do not foist your own views on others."

"I repeat: I am not addressing all of you, but only those who have not lost their human dignity and self-respect."

"Quite right, Tamara! Quite right!" shouted several of us.

Tanya Stankovskaya clumped past us in her oversized boots toward the door. "Let's demand it, then," she shouted, and, without further ado, began to hammer on the door with her bluish fists.

The train was slowing up for another stop. "Water . . ." she cried. Others joined in: "Swine! Torturers! You've no right to do this! You're subject to Soviet law the same as we are!" Someone else added: "We'll smash up the car! Shoot us and be damned! Wa-ter! . . ."

There was a clatter of feet on the platform; the doors were flung open with a jerk. The Brigand appeared with four of his men. His eyes were bloodshot as he shouted:

"Silence! Are you all out of your minds or something? What is this, a mutiny? Who are the ringleaders?"

As no one answered, he grabbed Tanya, who was nearest the door, and the inoffensive, silent Valya Streltsova, and ordered them to be taken off to the "punishment cell" car. Tamara stepped forward.

"We demand water," she said quietly. "All of us. The two you have arrested have done nothing wrong and Stankovskaya is very ill—she won't come out of the punishment cell alive."

Chava spoke even more softly and calmly. "We do not believe that it is part of Soviet justice to torture people by thirst. We regard this as an act of lawlessness on the warders' part, and we demand a proper ration of water."

"I'll teach you to demand things!" bellowed the Brigand,

in a voice half strangled by rage and astonishment. He was a very different person now from when he had listened almost like a normal human being to the Pushkin recitation.

"Mishchenko! Put the whole damn lot of them on short rations! And when we get where we're going, I'll show them what's what!" He made a vague gesture toward Tamara and Chava but, meeting their quiet glances, turned away after a moment's hesitation, preferring to treat as the ringleaders Tanya, who could hardly stand, and the silent, self-effacing Valya. The warders led these two off, but we did not calm down. Dozens of fists hammered on the walls and doors, repeating the infuriated cry of "Wa-ter!"

No one got up from the bunks any longer. The door was tightly shut and barred. Our bread ration had been cut to barely half, and we got no soup.

Yet we scarcely minded, or rather scarcely noticed, the Brigand's treatment of us. The one fear in all our minds was that Tanya would not come out of the punishment car alive.

Tamara was very dejected; she had almost stopped tossing her head in her usual manner. For three days in succession she declared to Mishchenko, when he brought the bread around, that there had been a mistake: it was not Tanya Stankovskaya but she, Tamara Varazashvili, who had had the idea of demanding a proper supply of water.

"And the other ringleader was not Streltsova, but me," said Chava Malyar quietly, her pale face still resembling Aïda's. "There are witnesses to prove it."

But Mishchenko had no time for these "educated" women and their twittering, high-faluting language. He merely growled phlegmatically:

"I can't tell, I wasn't there. Here, you, starosta, dole out the rations."

But Fisa Korkodinova, who had not been a Komsomol leader for nothing, decided to try to get around him. She stood stiffly to attention and said:

"Excuse me, Citizen Officer, may I have permission to ask you a question?"

Mishchenko was vastly flattered by her deference. He replied: "All right, but make it short, eh?"

"As the starosta, Citizen Officer, I should be glad to know how long you will be holding Stankovskaya, so that I can keep account. When will she be back?"

"Well, if it's to keep account, I can tell you—in two days' time. She'll be back the day after tomorrow."

But in fact she returned to us that very day. The Brigand obviously preferred to avoid the waste of time which making out a death certificate would have involved, and decided instead to deliver her alive to the transit camp at Vladivostok, where they could take what measures they liked.

Valya Streltsova, silent as usual, crawled back to her place on the middle bunk and did not even ask where her mug was. Nor did she thank Nadya Korolyova, who had carefully kept it for her.

(Eight years later, when Valya was dying of pneumonia she had caught during the haymaking season at Kolyma, where the prisoners slept in flimsy huts until November, we discovered the reason for her unbroken silence and withdrawal from people. The day before she died, she told her neighbor, a Seventh-Day Adventist named Arsenyeva, that during her investigation she had signed documents which resulted in several dozen death sentences for others. She had been a technical assistant to the first secretary of a Party regional committee, and she had been persuaded to denounce this man and a number of his colleagues. Arsenyeva was honestly convinced that it was her duty to tell this story to everyone in the camp after Valya's death: she believed that it would be a relief to Valya's soul if we knew that our sister, now standing before the Judgment Seat, had repented before she died and begged forgiveness of God and man.)

"Tanya," begged Pavochka, "do take my place on the bottom bunk. I can climb up quite easily, I'm young and healthy, but how can you in your condition?"

"It's all right," said Tanya hoarsely, "so long as I can stretch my legs out. In the punishment cell we had to sit with our knees bent all the time. Oh, it was a marvel of modern technique, that place."

We pulled off Tanya's boots and sprinkled a few drops of precious water on the edge of a towel brought from Yaroslavl, so that she could wipe the grime off her face. Musya, the doctor, felt her pulse.

"Why are your hands so cold?" I asked Tanya with alarm, having climbed up to visit her on the top bench. "It's so hot and stuffy here, and yet they're like ice. Have they got some means of lowering the temperature in the punishment cell?"

"No, it's even more stifling than here—not the slightest breath of air. I don't know what's the matter with my hands." She looked at them. They were like the crooked claws of an old, plucked rooster, lying across the counter in a butcher's shop.

(Some years later, when I was working as a nurse in the camp infirmary, I would come to recognize hands like these as a sure sign of a "goner" who was dying from undernourishment: so much so that whenever I came upon that clammy coldness I would, the same evening, start preparing a death certificate for the camp files.)

"Don't worry, Genia, I'm not going to die here. I can tell you as a friend, I want to die all right, but I must get to the transit camp first, I've no choice. After that we'll see."

"Because of your husband?"

"No—I'm not one of those optimistic fools. I know he's been shot. But I must get in touch with the men's compound—there's someone called Ivan Lukich from the Donbas, where I come from. He was made secretary of the district committee in 1935, and before that he made a name for himself as a Stakhanovite in the mines."

"Are you in love with him?"

"Don't be silly—he's sixty-two."

She coughed and wheezed for a long time before she could tell me the story. Musya had more than once said how worried she was by the state of Tanya's lungs, and when she listened to her cough she shook her head so hard that her black hair, tied with bits of rag, swung into her neighbors' faces.

When Tanya was arrested—she had previously worked in the cultural department of a district Party committee—the workers at the pit where her father and two brothers were employed had signed a collective appeal on her behalf. In it they declared that they knew the whole family from way back, that they had been mineworkers for generations, and that Tanya owed everything to the Soviet regime—a job, education, and a home. It was unthinkable that she should have anything to do with enemies of the state. It would have been like going against herself. Moreover, she was an excellent worker, and would stay on the job round the clock, if need be.

Over fifty of the miners signed this, and they took it to Ivan Lukich, who was then already secretary of the committee.

"Yes, he signed too. He didn't care what people thought.

He had sponsored me for Party membership back in 1922. He signed it, and he wrote in the margin: 'I vouched for her in 1922 and I vouch for her now.' "

"They were real friends, all right. What happened then?"

"Every single one who had signed, including Ivan Lukich of course, was put in jail too. I heard this when I was waiting for deportation, from people who came into prison after me. So now you see why I can't die till we get to the transit camp. The Social Revolutionary women, who should know, say we're bound to meet our menfolk there if they're still alive."

"You want to thank him, do you?"

"Thank him? I don't have to—he knows I'd do anything for him. No, I want to tell him that he shouldn't have done it—it didn't make sense. You see, I think that among the Party members who are still free there must be a lot like him, but for the time being they can't do anything."

"Why not?"

"I don't know. History will tell. But I do know that if they stand up against Stalin now no good will come of it, it'll only be death for a few more thousand people. A time will come for them to speak out, but why should they put their heads into the noose now when it won't do any good?"

Our conversation that night on the top bunk was interrupted by Chava Malyar. Her hair was disheveled and Tanya, on seeing her, stretched out her clawlike hand and sang—yes, sang—in a comic whisper: "Darest thou, Aïda, rival me?"

"Sh-h, Tanya. I've brought you . . ."

I rubbed my eyes. Was this a mirage? On Chava's open palm lay five lumps of sugar—brought all the way from

Yaroslavl, where we were given them at the rate of two a day; in the train we got none. It turned out that after her medical examination she had put one aside every day in case her heart troubled her. Sucking a lump of sugar was the quickest way to restore the pulse rate, and Tanya would soon feel better if she took two now and the others in the next few days.

(In spite of her weak heart, Chava Malyar lived to see happier times and even to read the records of the Twentieth and Twenty-second Party Congresses. With slow steps, peering about with her jet-black, apprehensive eyes, she climbed the stairs of the large building on the Old Square *; she also managed, before her death, to enjoy the luxury of hot running water in one of Moscow's new apartment buildings. In the spring of 1962 a small group of survivors of Car Number 7 followed her body to the crematorium.)

Sometimes, for reasons beyond our ken, the train would stop for days at a time. This was the worst torture. We suffered cruelly in the stinking, hot, motionless air. The door was shut tight and we were forbidden to talk, even though the train was in open country.

But at long last, we reached Sverdlovsk, where baths awaited us! The Menshevik Lucia Oganjanian, who had been there more than once in her time, told us nightly tales, as in the *Arabian Nights*, of the magical delights of the large, clean, spacious disinfection center in Sverdlovsk, just like the famous public baths in Moscow. There was a huge mirror in the changing room, everybody got a wash-cloth, one could wash to one's heart's content and even

* The Party headquarters in Moscow.

drink. We feared that the Brigand might deprive us of these pleasures as a punishment for the water riot, but in the upshot his wrath was short-lived.

"Starosta of Number 7," he bellowed. "Get your girls ready for the bath. That ought to stop their yelling. All the water you like to wash in, drink, splash about in . . ."

"Hope it'll make them pipe down," added the flat-nosed Mishchenko from behind his back.

Fisa Korkodinova came to life at the thought of such a large-scale operation. "Excuse me, Citizen Officer, how about laundry? When we left Yaroslavl they only gave us a jacket and a hand towel each. We've got no dresses or underwear except what we stand up in, and they're all filthy. . . ."

"Underwear can only be changed if it's badly stained by menstruation," declared the Brigand in a loud, official tone. "It will all be fumigated while you're washing, and then you'll have to put it on again. It may not look very nice, but at least it'll be disinfected."

"Well, now we've had our course in hygiene," sighed Nina Gviniashvili as the Brigand moved away.

"Girls! I've got an idea." Tanya Krupenik lowered her voice to a conspiratorial whisper. "Let's not give anything in to be disinfected except our jackets. We can wash our underwear under the shower and put it on again while it's still wet. In this heat it'll dry in no time, and at least we'll be clean."

"But the fiend's sure to find out," said Polya Shvyrkova sadly.

However, we were not in the mood to worry about laundry for the time being. What possessed our thoughts was the joy of realizing that our rough, sweaty, bony, neglected, tortured bodies would soon be in contact with water—hot water pouring from dozens of taps in the magic disinfection center at Sverdlovsk.

"All out! Form up in fives!"

The train had stopped on a siding where there was no platform. Mina Malskaya, Sofya Lotte, and several others could not face the jump to the ground. Pavochka Samoylova and Zoya Maznikova crossed their arms to make a support, and somehow the two were got down. But Zinaida Tulub, the writer, was still too frightened. Standing on the edge of the car, she explained at length to the guards, who stood by impatiently with their dogs, how bad Yaroslavl had been for her rheumatism and how athletic she had been in her youth, when such a jump would have been child's play to her. . . .

"Come on, woman, jump, damn it!" bellowed the Brigand. "Here, Mishchenko, give her a hand, can't you?"

Mishchenko dutifully presented his bull-like neck for Zinaida to hold on to, and in this way our distinguished though somewhat old-fashioned writer of historical novels was lowered to the ground. "I am really most obliged to you," she declared graciously to the panting Mishchenko, adjusting her Yezhov uniform.

Alongside each car there now stood a cluster of gray-brown, shadowy figures.

"Form up in fives! Close together now!"

The Brigand ran from one car to another, exchanging racy quips with the warders as he went. The German shepherds barked loudly and strained at their collars. They too were tired of the journey, and looked thin and mangy.

"I wonder what their daily ration of water is," said Nina Gviniashvili casually, to no one in particular. Mishchenko, who was standing close by, had no sense of irony; he answered quite cheerfully: "Oh, they get their bellyful—as much as they can drink."

By now the whole trainload of prisoners had been drawn up in fives, forming a wavy gray-brown line some seventy yards long. Some, overcome by the fresh air, slumped to the ground.

"Keep standing there! Hang on to one another! See that nobody falls!"

The guards, followed by their dogs, ran all over the place, vexed that some of us had fainted, in spite of orders to the contrary.

"Don't break the ranks there! Close up behind! Left march!"

"Are you sure," said Tanya Stankovskaya in a hoarse, broken voice, "that we shan't get another ten years for left-wing deviation?"

"Very funny, but she won't last long," Nina whispered in my ear. In the light of the early summer morning Tanya did indeed look like a goner. Her head drooped from her long, thin neck like a withered fruit from a branch.

"One step out of formation and we shoot," the guards warned us.

The changing room of the disinfection center surpassed our wildest hopes. It was indeed spacious and clean, and a mirror covered half the wall. Even so, it was not big enough for the several hundred naked women who jostled in front of it with washbowls in their hands. Hundreds of anxious, mournful eyes, all searching for their own reflection in the bluish glass . . . I recognized myself only by my resemblance to my mother. I called out to Pavochka: "Imagine, I'm more like Mother now than myself. How about you?"

She replied: "Without my hair, I'm more like Vanya."

Even in the excitement of looking into a mirror for the first time in three years, Pava could not for a minute forget her brother, to whom she was bound by great love and devotion.

(Pava, alone of our group, had the joy of meeting in Vladivostok the person whom she longed to see. Her brother Vanya was in the men's compound, separated from ours by a light fence through which she and he were able to talk. Vanya gave her as a memento a small pillow which he had somehow held on to, and kept saying: "Forgive me for ruining you, Pavochka."—"But how are you to blame?" she asked. "By being your brother," he replied. . . .

Before we embarked on the steamer *Dzhurma* which took us to Kolyma, Pava threw the pillow back to Vanya, as she had nothing else to give him. That was the last they saw of each other, for in 1944, while still in the camp, Vanya was denounced by an informer for "careless talk" and shot. His sister learned of this in the fifties, after she and her dead brother had been rehabilitated. . . .)

"What's the Brigand up to? Has he gone mad?"

The commander, with enviable nonchalance, was walking up and down among the hundreds of naked women. No sooner had we gasped at this than we noticed that two soldiers in full uniform, rifle in hand, were posted at each of the doors leading into the shower room.

"Saints above," wailed Polya Shvyrkova, like a peasant woman, "they must think we're not human, making us walk past the men with nothing on like this. Are they out of their minds or something?"

"Didn't your interrogators teach you in '37 that there are no differences of sex where spies, saboteurs, terrorists, and traitors are concerned? That being so, we needn't think of them as men either." And Nina Gviniashvili stepped boldly through the door between two soldiers.

"Yes," whispered Tanya Krupenik, "but these soldiers do see us as women and human beings. Just look at their faces.'·

She was right. Every single one of the sentries was staring at a spot between his feet, as though counting the pairs of heels that went past him. None of them raised their eyes to look. It was a different matter with the Brigand, who could not resist the pleasure of summoning to his presence the starosta of Car Number 7.

"Starosta of Number 7, stand before me like a blade of grass!" he bellowed, his mischievous, roving eyes glistening with anticipation. The naked Fisa duly stood before him. As she did so, a murmur of admiration spread through the crowd of women. None of us in the car had paid any special attention to Fisa's hair, which she wore combed back tightly behind her ears and coiled in a bun. But now her red hair fell freely in flowing locks, covering her body down to the knees. Standing there, with a bowl in her hand, she looked like some Lorelei of the Urals, or like St. Barbara, whose hair grew long by a miracle in order to shield her nakedness from the pagan torturers.

"Starosta of Car Number 7 reporting, Citizen Officer!" she announced in her deep voice, holding her hair across her breast as if it were a scarf she had thrown over her shoulders. The commander, with ill-concealed disappointment, gave her his orders, and the inmates of Number 7 crowded lovingly around their clever, quick-witted starosta.

Our blissful halt at Sverdlovsk lasted a whole hour. Our guardians had their hands full with the disinfection procedure and were not in a hurry. We were so cheerful that the splashing of water was accompanied by ripples of laughter. Tanya Krupenik, rubbing away at her underwear, even began to hum a song about the banks of the Dniepr. This girl's goodness and love of life were inexhaustible. She did not appear downcast for a moment at the thought that she, alone in Car Number 7, had been sentenced to twenty years and not ten. This, she believed, had happened because she had been tried on October 5, 1937, on which date a new law had come into force raising the maximum term of imprisonment from ten years to twenty-five. But many of the women whispered that it was because she was a close relative of Lyubchenko, the former president of the Council of People's Commissars of the Ukraine. As we all knew, the closer you had been to prominent Communists, the longer your sentence.

Anyway, Tanya laughed off the difference on the ground that no one would stay in prison as long as ten or twenty years—the Party would see to that. "Yezhov's been kicked out already, and the others' time will come. It's obvious that saboteurs have infiltrated the NKVD—they're sure to be unmasked, and then we shall be let out. Even now, things are better than in Yezhov's time: he kept us two years in solitary, but now we're getting a chance to work and develop the Far North, which shows they know we're good workers."

(In 1947 Tanya discovered the difference between a ten- and a twenty-year sentence, when those of her companions who were still alive were transferred, one by one, from the camp to a settlement,* their places being taken by people arrested during the war. Tanya, who despite prisons and camps had retained her outlook on life as an old Party member of the twenties and thirties, felt a

---

* People released from camps were often required to remain as "free workers" in remote areas, generally in the vicinity of the camps.

stranger among the newcomers. . . . In 1948 a fire broke out on the collective farm at Elgen in the Kolyma area, and Tanya, who was employed as an agronomist, found herself threatened with a fresh trial and a sentence for arson and sabotage. So, on one of the white nights of the short Kolyma summer, the brown-eyed, black-browed Tanya was found hanging from a noose in one of the greenhouses where cucumbers and tomatoes were grown for the camp officials. A swarm of Kolyma mosquitoes buzzed about her head—bloated, repulsive insects that reminded one of small bats.)

During the first day or two out of Sverdlovsk, we felt much more cheerful. We began again to recite poetry and to give talks about subjects in which we were specialists. Zinaida Tulub gave us from memory passages of Maupassant, in French. These were greatly admired, and even Lena Kruchinina forgot her indignation about Zinaida's cat Lyric.

Mina Malskaya stopped clutching at her heart and gave us—not a lecture, this would have been too much in her condition—but a brief summary of the geography of the Far North.

But on the third day we found that our water ration seemed even more miserable after the luxuries of Sverdlovsk. Quarrels and listless bickering again became the order of the day.

"Oh dear, oh dear," sighed Polya Shvyrkova, "what a pity one's belly has no memory! I must have drunk at least a gallon at Sverdlovsk, and now—what a state of affairs!"

"Oh, stop whining and making a noise," said one of the lucky ones who were able to sleep around the clock.

"Don't quarrel, girls!" sighed Nadya Korolyova. "I can't understand how people can get so mean!"

"I can," said Tanya Stankovskaya hoarsely. "They cast their nets very wide and they caught all kinds of fish. . . ." She no longer stirred from her upper bunk, and when I asked how she felt she would merely answer: "I'll get to the transit camp, don't worry!"

One day at dawn, not far from Irkutsk, we were wakened by a violent jolt.

"Was that a collision?"

"I wish we could have one that'd smash up this old car. Then they'd have to let us out into the fresh air and we could breathe for a change."

"No such luck. Hey, look out, the mugs'll fall over." This was a calamity far more serious than any collision—without our mugs we would never hold out till Vladivostok.

The train stopped. Outside we could hear the guards running from one car to another, shouting and cursing.

"What do you think it can be?" asked Zinaida politely for the hundredth time, looking entreatingly around the car with her dark, dreamy eyes which made her look more like Pushkin's friend, Anna Kern, than an inmate of Car 7. . . . No one replied, till after a while Tanya's voice was heard from above:

"It's either Vesuvius erupting or it's Stalin replying to Olga's poem."

Nadya's mug had broken in two as a result of the jolting, and she was sobbing as over a dead child.

Suddenly our door was wrenched open so violently that we jumped. Evidently this was not an ordinary visit of the guards. What could it be? First the commander, then a crowd of women, pushed from behind by two of the guards—about fifteen of them, so weak that they leaned against the car walls, and all dressed in Yezhov uniforms.

"Starosta of Number 7, here are some reinforcements for you!" shouted the Brigand. "Move up a bit and make room, you're sprawling about like ladies-in-waiting."

"But, Citizen Officer, we can hardly move as it is," said Fisa, protesting for once. "We can't even turn on our sides unless we do it all together."

"Well, you should keep still. What do you keep tossing and turning for?" quipped the commander, while Mishchenko added: "Yes, lie down and don't fidget."

Nadya, still sobbing, pushed her way forward to tell her sad story. It wasn't her fault there'd been a collision, and how was she to manage without a mug?

"Well, you won't get another—I have to account for them, you know. There's a full stock-taking in Vladivostok. It's the citizen's duty to look after state property," the Brigand added sententiously. Then he and his men jumped down, one by one, onto the sandy ground and the door was barred again.

The new arrivals huddled together near the hole used as a latrine, clutching their coats to their breasts. For a few moments there was dead silence. Some of the original inmates of the car looked far from friendly. As it was, we were barely able to keep alive, with no air and nothing to drink. How could we possibly find room for a new batch of people?

"Here, who's been shearing you like that?" It was Polya, who first noticed that there was something strange about the newcomers' appearance compared to ours—something even more degrading and unbearable.

"Yes, they shaved our hair off. We're not from Yaroslavl. We were all in the Suzdal prison, and we only went to Yaroslavl to be put on the train there. Our car—Number 12—was smashed in the collision, so they've split us up into three groups and moved each of them into a different car. But you seem to be pretty crowded already."

Suzdal . . . the second-biggest solitary-confinement prison for women. When we were at Butyrki, we had hoped to get sent there because it was a former convent and the nuns' cells were bound to be drier. And now here were the Suzdal prisoners with their heads clean-shaven. We looked at them with horror, and they gazed with envious admiration at our gray, bedraggled, dusty braids, curls, and bangs.

"Oh, you poor dears!" exclaimed Polya. The words were like a signal. All of a sudden we saw in the newcomers not parasites with whom we had to share our meager ration of water and air, but our own sisters, humiliated and suffering even more grievously than ourselves. Their hair—fancy shaving off their hair!

"Come over here, comrade! We can move up a bit."

"Put your coat on top of mine."

"Take your boots off and put your feet up here. We can manage, there's not so far to go now—we're in Siberia already."

One of the newcomers recognized me and pushed her way through to me with the help of her rolled-up coat, crying "Genia!" I did not at once realize that it was Lena Solovyova, whom I had known well in the old days in Moscow. It was hard to connect the laughing, flirtatious Lena with this almost sexless figure who looked as if she had just recovered from typhoid fever. A whitish stubble was beginning to sprout from the top of her grotesquely large skull, and her prison dress hung from her shoulders as though hanging from nails. . . . From my long, blank stare and the tone in which I finally exclaimed "Lena darling!" she must have realized for the first time since her imprisonment what had become of the cheerful, clever postgraduate student that she had once been. Hastily, she pulled a dirty, crumpled scarf out of her pocket and threw it over her head, saying: "Well, do I look a little more like myself now?" Then she stretched out her bony hand with its swollen joints and exclaimed: "You lucky thing! You've still got the same curls as you had in Moscow."

I felt a spasm of acute, unbearable pity. It was, after all, only 1939, and in spite of the investigation, my "trial," Butyrki, Lefortovo, and Yaroslavl, I still had a lot to learn about the treatment that could be meted out to human beings. Thus I felt that the shaving of the heads of the Suzdal prisoners, and especially of my old friend Lena, was the supreme insult to their femininity. . . . Two or three years later I paid no attention to the appearance of any woman's head, covered as they all were by the caps which made the inmates of the camps look like helmeted Pecheneg warriors.

"Your hair'll grow again, Lena, it will, and you'll be as pretty as before. Don't envy us—we're in the same boat as you are, and they might do the same to us." I felt my own hair with my hand. No, that was something I believed I could scarcely survive. . . .

"Lena, what's happened to your Ivan? And the girls?"

Her masklike face quivered. "I don't know anything about the girls, I wasn't allowed to write. And Ivan is where all decent people are."

She said this matter-of-factly, as one who had nothing left to fear and did not care if her words were reported by some orthodox Stalinist who happened to overhear them.

Among the Suzdal women there was one whose head had not been shaved. "I would not permit it," she explained in a firm, precise tone from which I immediately deduced that she had been a teacher. Her name was Lila Itz and she was a headmistress from Stalingrad. She was tall and handsome, with auburn curls which fell abundantly over her shoulders. Her pupils had worshiped her, but they also nicknamed her "Lila the Terrible." "I snatched the scissors, hit the barber as hard as I could, and bit his hands," she went on in the same schoolmistress's tone. "I saved my hair, but I got my leg hurt instead—which is less of a nervous shock." Her right knee was black and blue; her leg looked like a log and was all shiny. "They threw me into the punishment cell as hard as they could, and I went right into an iron bunk. But they didn't cut my hair off, because the convoy was about to leave and they had no time to go on arguing with me."

Tamara Varazashvili pressed her hand warmly and said: "Comrade, I admire your courage." From the corner in which our professional Marxists had ensconced themselves (and of which they were not giving up a single square inch) came a sound of dissent.

"Didn't it occur to you," asked Lena Kruchinina, "that the shaving might be for purely hygienic reasons, as a precaution against lice, for instance?"

The women from Suzdal, one and all, rejected this theory. They had obviously discussed the matter at length among themselves.

"How could lice have got into our cells? They were bare and clean—nothing in each of them but stone, iron, and one person in prison uniform. Twice a month we had a shower, each of us separately. No, it had nothing to do with hygiene, they just wanted to insult us."

"Well, just to crop someone's hair is hardly an insult.

Not like a Tsarist prison, where they'd have left you half shaven."

This was more than Tanya Stankovskaya could stand. Somehow or other, she found the strength to scream so that the whole car could hear: "That's the spirit, girls! A vote of thanks to Comrade Stalin! Life has become better and gayer,* he said, and to prove it we have all our hair cut off instead of only half. Let us give thanks to our father, leader, and creator for our happy lives!"

"Stankovskaya, to hear your anti-Soviet talk, one can hardly believe that you were a member of a municipal committee!"

"Yes, and to hear you people one can hardly believe that you're not on the prison staff. Why don't you call the guards now and report this conversation? You might get some clean underwear as a reward, and then you wouldn't stink so much."

"Sh-h, girls, what's the use of talking like this?" said the naïvely orthodox Nadya Korolyova imploringly. "It isn't nice to insult one another. Look at me—I've broken my mug, but I don't lose my temper with everybody. We must just put up with things—prison is prison, not a health resort. Remember the old prisoners' ballad:

Here are we, the Emperor's guests,
As befits our station. . . ."

The wheels rattled more and more slowly and irregularly. Surely we could walk to Vladivostok faster than this! Car Number 7 was overworked and overloaded, but it crawled on and on across the Siberian wastes. Drowning

* Famous saying from a speech of Stalin's.

the sound of the wheels, we took up the refrain of the convicts' song:

> A curse on our unlucky fate,
> Doomed to transporta-a-tion.

Some of the Suzdal women had quite a reputation: for instance Lina Kholodova, who had been a machine-gunner in the civil war. A rumor had gone through the convoy that she was Anka, who had fought under Chapayev, but she was in fact Lina and had been with Shchors' troops.* The investigators had said to her: "Weren't there enough men in your own village, that you had to go whoring all over the battle front?"

Then there was a parachutist named Klava Shakht. After two years of Suzdal and prison uniform her deportment and manners were still magnificent. Only her fingers were deformed, because on her last jump she caught them in some wires.

Felya Olshevskaya, who had been a Party member since 1917, had worked for many years with the Polish underground, and her sister was married to Bierut.†

Tadjikhon Shadieva was a collective-farm manager from Uzbekistan. Her face seemed familiar to many of us because in the thirties it was frequently seen in film documentaries and on the covers of magazines like *Ogonyok* and *Searchlight*. After three years in prison she had still not got used to the idea that it was no longer her business to worry about the nation's economy. She kept on telling us with enthusiasm and in poor Russian about long-ago

---

* Vasily Chapayev, 1887–1919, and Nikolay Shchors, 1895–1919, famous partisan leaders of the Bolsheviks in the civil war.

† Boleslav Bierut, 1892–1956, Polish Communist leader who became Prime Minister and head of the Polish Communist Party after the war.

plenums in which she had discomfited a certain Biktagirov in connection with delivery dates for cotton. With Tanya Krupenik and Anya Shilova, she discussed at length ideas and plans for the agricultural transformation of the Kolyma region.

I could not stop looking at Lena Solovyova and rejoicing that we had met again. She had known my elder son and I knew her daughters. In 1936 we had sat side by side at a translators' conference in Moscow, and we reminded each other now of the interesting addresses given by Pasternak, Babel, and Anna Radlova. I stroked the stubble of whitish hairs which covered the top of Lena's skull, and from time to time tore myself away from the conversation in order to climb up and see how Tanya Stankovskaya was getting on.

"It's all right, don't worry, I'm still alive," she invariably replied.

(Sure enough, Tanya did not die before we got to the transit camp. But she was not able to look for her Party sponsor, Ivan Lukich, for when we arrived she was no longer capable of walking. She was carried into the camp and placed on a bunk from which she never got up again. There she lay, almost fleshless, more like her own shadow than a human being. Not even the bugs went near her—those bugs at the Vladivostok transit camp behaved almost as if they were rational, organized beings, converging purposefully in massed attack on each new batch of prisoners. Just to make matters worse she developed night blindness, and could not see me when I came and held her hand.

On the day of her death, a rumor was going around that Bruno Jasienski * was somewhere in the camp and had died

* Polish Communist writer, 1901–1941, who emigrated to the USSR in 1929, where he was very popular. Arrested in 1937.

that day of malnutrition. When I told Tanya this she laughed, baring her teeth, which stuck out hideously in all directions because of scurvy, and said: "What luck! When people remember me, they'll say 'She died on the same day as Bruno Jasienski, and of the same disease.'"

These were the last words of Tanya Stankovskaya—a mining girl born and bred, the Party "godchild" of Ivan Lukich, and the bravest inmate of Car Number 7.)

As Vladivostok drew nearer, the discussion turned more and more to the subject of footwear. It seemed that we were to get off the train not in the town itself but at a place called Black Creek. From there we should have to walk several miles to the transit camp.

"How can we walk in these boots? They fit so loosely, our feet'll be raw with blisters."

"Suppose we ask the Brigand for footcloths?"

"Where will he get them?"

"I'd like to see Vulturidze made to tramp in these for the rest of his life!"

But he had scarcely invented them—they were a product of the ingenious fancy of Comrade Yezhov, Stalin's People's Commissar, the nation's favorite son. . . . However, as there were not enough prison boots to go around, some of the women were still wearing the shoes they had been arrested in—the heels broken, the soles falling off and tied with string.

The guards' restiveness also suggested that our journey was coming to an end. Several times a day they counted and recounted us, drew up lists and copied them out. Mishchenko in particular had his hands full. He had to make a separate list of the Suzdal women in Car 7. He had attempted this exacting task three times, each time postponing it to the next day.

"What is your alias?" he asked Tadjikhon Shadieva.

"I'm not a common criminal to have an alias," she retorted. "You'll have to make do with my surname."

"Well, what is it?"

"Shadieva."

"Your nationals?"

"Uzbek."

"No, not nation, nationals."

"He means your initials," explained the quick-witted Musya.

"Oh, I see. T.A."

"In full, in full!" bellowed the baffled Mishchenko. . . . He also had fearful trouble with German names:

"Gat-zen-bul-ler . . . Tau-ben-ber-ger . . ." Then, wiping the cold sweat from his forehead: "Nationals?"

"Charlotte Ferdinandovna."

No, copying out jawbreaking names was no job for an honest peasant. . . .

We reached Vladivostok exactly a month after leaving Yaroslavl. Or rather, not Vladivostok itself but a deserted place in its environs, which may or may not have been Black Creek. It was night when the train stopped. Outside, a reinforced team of guards was waiting to take delivery. The German shepherds, straining at their leads, made a terrific din.

"Everyone out! Form up in ranks of five!"

Suddenly we could smell the sea air. I felt an almost irresistible desire to lie flat on the earth, spread out my arms, and disappear, dissolve into this deep-blue space with its tang of iodine.

Suddenly despairing cries were heard:

"I can't see! I can't see anything! What's the matter with my eyes?"

"Girls—please give me a hand. I can't see a thing. What's happened?"

"Help, help, I've gone blind!"

It was night blindness, by which about a third of us were affected immediately we set foot on Far Eastern soil. From dusk to dawn they could see nothing and would wander about, stretching out their hands and calling to their comrades for help.

The panic of those affected now spread to all of us. The guards furiously demanded silence so that they could count the heads they were handing over. Luckily, Musya came to the rescue with the medical knowledge she had acquired barely three years earlier.

"Don't be afraid, girls," she cried, shaking her braids and wiping away a stream of tears with her sleeve. "Listen! You haven't gone blind, it's only night blindness due to lack of vitamins in prison and on the journey, and to low resistance generally. It's touched off by the sudden change of climate. It's curable, do you hear? And it only lasts from dusk till dawn. You need cod-liver oil for it—three spoonfuls is enough. Don't worry, my dears!"

The hand-over of prisoners took the rest of the night. Dawn brought to the horizon shades of mauve and purple that we had never seen before, and a bright-yellow sun that looked like something drawn by an artist's hand.

"At last I shall understand Japanese painting properly," said Nina Gviniashvili, gazing at the sky as she wound around her feet a piece of toweling from Yaroslavl, doubled over so as to ease the chafing of her boots on the long walk to the transit camp.

Once again the long, wavering, gray-brown line. . . . "Quick march, now! Not too fast in front! Watch out, we'll shoot if you break formation."

We lumbered off in our clumsy boots, which sank into the sand at each step. I looked back. There, in the rays of a magnificent picture-postcard sun, I caught a view of the old, dirty, dark-red freight car, which looked all warped. Across it, in bold chalk letters, was written: "Special equipment."

3

• *The transit camp*

And so we trudged on and on. It was the dawn of July 7, 1939, and gave promise of being a scorching day. But for the moment we felt on our faces a breeze smelling of newly washed laundry. We breathed it in greedily, feeling that we were getting rid of the grime of Car Number 7. Our road rose and fell, and from start to finish we did not meet a living soul: not a man or woman, not a horse or a vehicle. It was as if the world had come to an end, and we, its only survivors, were living out the last days of an unbearable life.

Marching along, I felt as if I were asleep and seeing this empty road that smelled of the sea in a dream. Only my companions' shouts and groans brought me back to real life. The shouts, oddly enough, were of joy: those who had been blind the night before were crying out: "I can see!" Some of them had not grasped that when evening came they would go blind again.

The Suzdal women were far weaker than we from Yaroslavl. With their shaven heads, they all looked alike, as though mass-produced in a horror factory.

The road was endless. I do not know to this day how many miles it was. The guards shouted themselves hoarse, the German shepherds yelped wearily like harmless mongrels. It got hotter and hotter. At all costs one must not fall down—for ahead of us lay the transit camp, that longed-for place where the men's and women's compounds were separated only by barbed wire and where we would meet our menfolk—even my husband might be there. . . . With ready credulity we believed this legend, based on the outdated experience of the Social Revolutionaries and Mensheviks among us. This crazy hope sustained us, half-living ghosts that we were, as our march continued up and down hill under the ever fiercer sun of the Maritime Province.

At last we reached the gates of the transit camp, which were protected by a dense barbed-wire entanglement. The dogs barked more cheerfully, sensing that their mission was about to come to a close.

"Through the gate in ranks of five!" shouted the guards, pushing forward those of us who were on the point of collapse.

Inside the barbed-wire fence we saw women, hundreds of women, wearing faded, patched, ragged dresses and blouses, with emaciated, deeply sunburnt faces. These were camp prisoners: they had not felt the deadly breath of the cells of Yaroslavl and Suzdal. They looked no worse than a crowd of beggars, refugees, bombed-out people, whereas we were figures from a nightmare. This could be seen in their faces as they looked at us with an expression of horror, and in their readiness to share their last rags with us. Many of them wept openly as they watched us trudge through the gate in an interminable gray stream. We heard their whispered comments:

"Look at those convict stripes—the Yezhov uniform."

"People from the prisons—two years and more in solitary."

"People from the prisons"—for the next ten years or so this grim definition stuck to us like a label. We were the worst criminals, the worst off, the worst everything. . . .

We did not realize at once the full horror of our situation. It took time to discover that, whereas at Yaroslavl we had all been more or less equal, there was no equality in this new circle of Dante's Inferno. The camp population was divided by the devilish ingenuity of our torturers into numerous "classes." To begin with, the aristocracy consisted of people who had got into trouble for such respectable crimes as embezzlement, bribe-taking, and so forth. (We did not come into contact with the more "vulgar" sort of non-political criminals until after we left the transit camp.) The embezzlers and suchlike were very proud of the fact that they were not "enemies of the people" but were making good their misdeeds by honest work. It was they who were appointed to almost all the positions of authority—as foremen, orderlies, starostas, and so forth—open to prisoners.

Then came the complicated hierarchy of the Article 58 people, the "politicals." The most innocuous group of these fell under section -10: the "babblers," retailers of political jokes, and so on, or, in the official phrase, anti-Soviet agitators. Next came the CRA's—those convicted of "counter-revolutionary activity." These were mostly not Party members and got off with lighter work or even administrative duties. The SE's ("suspected espionage") were sometimes lucky enough to get into this category also.

Next came the CRTA's ("counter-revolutionary Trotsky-ist activity"), who until our arrival were the lowest category of all, the camp pariahs. They were assigned to the hardest outdoor work, were never appointed to any of the jobs mentioned above, and from time to time were put in punishment cells on state holidays.

To those in the last category, our appearance was a great consolation. Their crimes were as nothing beside ours, which had merited condemnation by a military tribunal to terms of imprisonment under the articles concerned with terrorism, and we were also a mighty reinforcement for the heavy work of tree-felling, land-improvement, hay-making, and so on.

In point of fact, the main difference between us and the CRTA's lay in the date of our arrest. They, like us, were mostly Party members, but they were arrested at a time when the penalty for political offenses was usually not more than five years, whereas the victims of Yezhov and Beria were condemned first to ten and later to twenty or twenty-five years. Paradoxically, those arrested earlier had for the most part some real opposition background, having at some time or other failed to vote on the right side, or abstained, whereas we were mostly orthodox Communists, members of the intelligentsia and the Party machine. But no one, of course, bothered about this incongruity.

A current joke in the transit camp was that whereas German women's lives were supposed to be bounded by four K's—*Kinder, Küche, Kirche,* and *Kleider*—ours could be summed up in four T's: Trotskyism, terrorism, toil, and torment. . . . Another joke related to the medical examination. The doctor, stethoscope in hand, says "Breathe" and then asks: "Which article do you come under?" On being told "Article 58—ten years" he says: "Oh, well, in that case don't breathe."

Indeed, there was not much to be gained by breathing. With predictable cynicism, the camp medical system took strict account of the victim's sentence. People in our category were automatically classed as "fit for hard labor"— not excluding Tanya Stankovskaya, four hours before she died. . . . This was our first contact with camp medicine, and it taught us new things about the doctor's profession. On the one hand, even a prisoner of our group who happened to be a doctor was almost certain to be employed as such and thus to save her own skin. On the other hand, it was harder for doctors than for anyone else to keep a clear conscience and avoid selling for a mess of pottage the lives of thousands of their comrades. They were tempted every minute of the day by the offer of a warm corner in the "service block," bits of meat in their soup, or clean quilted coats of good quality. We were not to find out until we reached Kolyma which of them would or would not succumb to these temptations. But we noticed at once, for instance, that when Anya Ponizovskaya from Suzdal became a member of the medical board, with a white overall covering her prison garb and a Red Cross scarf on her shaven head, she immediately ceased to stoop and her voice took on a somewhat metallic note, which, at least as yet, did not grate on the ear.

The camp consisted of a huge, dirty enclosure surrounded by barbed wire and reeking of ammonia and the chloride of lime which was constantly being poured into the latrines. I have already mentioned the special breed of bugs which infested the long wooden hut, with its three tiers of bunks, in which we were housed. For the first time in my life I saw these insects behave in an organized, almost conscious manner, like ants. Instead of crawling, they moved swiftly in large, impudent, purposeful groups, already gorged with the blood of previous arrivals. One could not sit on the bunks, let alone sleep, and from the very first night there was a mass migration out of doors. The lucky ones managed to find boards, bits of broken fences, or mats; the less resourceful had to spread their Yaroslavl jackets on the dry Far Eastern ground.

I shall never forget the first night spent in camp in the open air. Since my two years in prison it was the first time I had slept under the stars. A fresh breeze blew from the sea and brought with it an illusory sense of freedom. The constellations wheeling overhead seemed from time to time to change their shapes. I remembered Pasternak's words:

> The wind, warm and selfless, caressed the stars
> With something of its own, eternal and creative. . . .

The lilt of the words seemed to be repeated in the singing of the telegraph wires. The gusts of sea air finally overcame the acrid smell of chlorine. By an effort I succeeded in feeling almost happy. It was a long time till morning, when life would begin again, and meanwhile I had the stars, and poetry, and the sea close by. I reminded myself that Vladivostok was a port, from which steamers sailed every day to distant, unknown lands. Couldn't I pretend that our camp itself was such a steamer, a temporary abode till we reached our destination? . . . Closing my eyes, I surrendered my senses to the intoxicating reunion with nature after a long separation. Where would our Noah's ark come to rest? As the stars twinkled overhead, so our lives glimmered, flickering and uncertain as sparks borne by the wind. . . .

At dawn, when colors and shades began to be visible, I discovered with delight in the corners of the yard scattered nettle bushes and large, dusty burdocks. I was staggered by their green magnificence. The nettles, sting as they might, were so beautiful, and the burdocks so kind and trusting. They had come all the way here from that far-off land where I had spent my childhood—from a back yard in a lane off Arbat Street in Moscow. . . .

"Get up! Breakfast!"

It arrived in mobile kitchens, before which we formed up in long queues. There were very hot soup and some deadly "pies," which very soon doubled the already large numbers of those who were suffering from diarrhea.

During the first three days, while the medical inspection was in progress, we were not sent out to work and spent most of the time in making new acquaintances. With the curiosity of ex-"solitaries" we talked incessantly to the camp women, many of whom had been here for more than a month. One after another, we learned their life stories—all of them fantastically improbable, and yet true; all tragic, yet consisting of episodes which were comic in their incongruity.

"I'm sure we have met somewhere, comrade," said Sofya Mezhlauk, the wife of Molotov's deputy, staring at Tanya Krupenik.

"Yes, I remember your face too. We must have met either at Butyrki or in the government rest home."

We did our best to retrace the course of events, starting from the moment at which we were arrested. For the thousandth time we heard variations on the theme of the Great Leader who devoured his people. The camp women knew much more than we did, because they had mostly been arrested later and, being far less strictly treated than we, they had some opportunity during their work to meet people outside the camp.

On the day when Tanya Stankovskaya died and I was

sitting sadly by myself, a young girl with a pleasant face that reminded me of a sound, rosy apple came up and said quietly:

"You mustn't grieve so much for your friend. People die here so often, you can't afford to. Think of something else, your family for instance. Have you got anyone outside?"

"Yes, my children and parents. My husband's been arrested."

"There you are, then. I work outside the camp—if you write a letter, I'll post it for you."

Imagine—I could actually write Mother a letter without the interference of the Yaroslavl censor! I hastened to cover two sheets of the girl's notebook with tiny writing, so as to get in as much as possible. The notebook itself, which she drew so casually from her pocket, seemed to me as miraculous as if it had been a handful of diamonds. My delight and astonishment knew no bounds when, with the same casualness, she gave me a real stamped envelope. I could not believe my good fortune, and handed her my letter with the sort of feeling that shipwrecked sailors must have when they launch into the sea a bottle containing a call for help.

This girl, whose name was Allochka Tokareva (CRA, ten years), took a fancy to me and acted as my good fairy during the whole month that I spent in the transit camp, instructing me kindly and tactfully in the conditions of my new life.

"When they start filling up forms about you," she told me, "say that before you studied arts you had a year or two of medical training. Then, at Kolyma, you've got some chance of being made a nurse instead of having to dig the ground or cut down trees."

"But it's not true! How can I be a nurse?"

"Of course you can! The main thing people need is to have a decent person looking after them. You'd do your best for the 'goners,' and you wouldn't take bribes."

"But what about treatment?"

"Don't be silly! The only treatment they get is to be let off work for a day or two."

"Well, I can't tell lies. . . ."

"Then you'd better learn, hadn't you?"

These words on the lips of a young, apple-cheeked girl struck me as yet another consequence of the Great Lunacy.

At the outset, being slow to learn, I made an enemy once and for all of a certain Tamara, a powerful person with the imposing title of team leader. She was, in fact, the first example known to me of a criminal occupying a "soft job" at the camp. She had managed this despite being an Article 58 prisoner and, I think, a CRTA, and this fact, if I had known it, ought to have put me on my guard. As it was, hearing only that she was a former Komsomol official from Odessa, I asked her innocently whether any of our things had arrived from Yaroslavl. This was a question of some urgency for me, as the red slippers I had worn out pacing about my cell there were now useless and, as I had not been issued with boots, I was virtually barefoot. My old black shoes, which were still quite serviceable, were among the possessions taken from me at Yaroslavl, and I longed for them so much that I saw them in my dreams.

I put my question to Tamara as politely as I could, calling her "comrade," as was the custom among us "people from prison." For reply she turned to her underling, a non-political who was always at her side, and said: "What sort of mother's darling have we got here?" Her handsome, regular face, which was normally conspicuous for its pinkness among our sallow gray ones, was bright red with anger. I learned afterward that she was one of those in her position who were constantly on the lookout for an opportunity to let fly at someone. She went on:

"I'm sorry that Madam's trunks have not arrived yet. Perhaps by the next courier? . . . And don't call me comrade. If you want to ask questions about your precious things, go to your starosta and don't come to me." She screamed all this in a loud falsetto, banging the table and drawing everyone's attention to us. Then she went on another five minutes or so, berating me for lack of deference.

I made matters worse for myself by retorting: "I beg your pardon, I did make a mistake—you are not a comrade." By these words I made a mortal enemy of this powerful person, and it was not long before I felt the consequences. Three days later, when I could still hardly stand as a result of scurvy, loss of weight, and diarrhea, I was sent out to work in a stone quarry.

Some of the women from Odessa who had known Tamara in former days said that she had been an excellent girl, kind and friendly, and a first-class Komsomol. Afterward I came across many such cases of a complete change of personality brought about by the struggle for life in camp. In some people, the former self seemed to have been completely obliterated and replaced by some frightful monster or a robot without a soul, without human feelings, and above all without memory. Such individuals seemed to have no recollections of the period of their lives before they were arrested—indeed, they could not afford to have them.

The Odessa women knew this well and never approached

Tamara as former comrades of hers. By keeping haughtily
to herself she insulated herself both from condemnation of
her own behavior and particularly from the recollection
of what had happened to her. Her constant bad temper and
the ease with which she both took and gave offense, insulting those in her power, were due to contempt for people
and a secret fear of them. She treated her numerous toadies
with indulgent scorn, and hated and persecuted those who
showed by silent looks that they understood the workings
of her mind and character.

By innocently calling her "comrade" I had reminded her
of the past which she had cast out of her mind because it
stood in the way of her present career: hence the outburst
with which she answered me. After the incident I started
to reflect on the psychological type created by camp conditions, and whenever I met Tamara afterward I was reminded of Blok's lines:

> How terrible to be a corpse among the living,
> Pretending to be alive and full of feeling!
> But why pretend? To be accepted by society,
> One needs only to conceal the rattling of one's bones.

In later years, in the camps, I met many of these spiritu-
ally dead people. In prison there were none. Prison, and
especially solitary confinement, ennobled and purified human beings, bringing to the surface their finest qualities,
however deeply hidden.

My first acquaintance with hard labor was made in a
stone quarry. It was a blazing hot July, under the merciless, ultra-violet rays of the Far Eastern sun. Even at a
distance, one could feel the infernal heat given off by the
stones. And we, who had not seen a ray of sunlight for
over two years, who had become unused to any kind of

physical work, who suffered from scurvy and pellagra and the effects of the journey in Car Number 7—we were now to perform labor in the fields and quarries which would have required all the strength and endurance of grown men.

It is astonishing how few of us got sunstroke in that broiling heat. One was Lisa Sheveleva, a Comintern employee who had been private secretary to Elena Stasova *: She died in the infirmary at Magadan immediately after we arrived there by sea. The other was Veriko Dumbadze, a girl arrested on her father's account when she was sixteen. Both of them were taken from the quarry to hospital (in the same barrows that were used for carting stones) but after two days they were passed as fit for work again.

Various literary reminiscences kept going through my head. New Caledonia, Jean Valjean, a convict fettered to a wheelbarrow . . . So this was what it was like in real life!

The trusties in the transit camp kept telling us that this work was sheer paradise considering the date of our arrest and the articles under which we were convicted. For instance, we had no norm to fulfill and were receiving a full ration of food. At Kolyma things would be very different. This was just a place to rest while we waited for a boat to take us over. But in spite of such "liberalism," many of us were suffering from ulcers that were due to malnutrition. At night, even in the open air, it was hard to sleep owing to the groans, cries, and labored breathing of hundreds of one's neighbors. By day, we constantly felt dust from the quarries between our teeth.

* Born 1873. Old Bolshevik who occupied a prominent position in the Comintern. Since Stalin's death she has written her memoirs.

Many people also developed a sort of mental torpor which enabled them to contemplate with indifference the dying, those afflicted with night blindness, wandering about in the evening, with companions to guide them and stretching out trembling hands, and even the hordes of bugs which crawled all over the plank beds. Others even got into the frightful beggar's habit of displaying for show their sores and the ragged remains of their Yezhov uniforms.

But these were the minority: by far the greater number actively clung to life. We still took pleasure in the fugitive mists of morning, the violet sunsets that blazed over us as we returned from the quarry, the proximity of ocean-going ships which we felt by some sixth sense—and in poetry, which we still repeated to one another at night. We were still living in that bitter-sweet world of feeling which our prison life in Yaroslavl had created for us. I felt instinctively that as long as I could be stirred to emotion by the sea breeze, by the brilliance of the stars, and by poetry, I would still be alive, however much my legs might tremble and my back bend under the load of burning stones. It was by preserving all these treasures in our minds that we should resist the onslaught of the horrors around us.

Some of us were already beginning to miss our solitary cells.

"I really think we were better off there. It was clean, and we had books, and we didn't have to work like beasts of burden."

"If we had stayed there, we'd have died of scurvy within a year."

"Well, what about now? Do you expect to live till next year?"

One day at dawn, when the pale stars were still twinkling between the clouds, we were aroused by a Social Revolutionary woman called Nadya Lobytsina. Although only thirty years old, Nadya seemed to us a living anachronism. Her spectacles, old-fashioned hair-do, and way of speaking all reminded one of the woman student of the early 1900's. But at this moment she was behaving like a scout or trapper straight out of Mayne Reid. She placed her ear to the ground and listened, holding one finger in the air. Then she jumped up from the board on which she had been sleeping and announced in a stage whisper:

"They're coming! A huge convoy—I'm sure it's the men at last! What did we tell you?"

Several of us looked at her pityingly. Was she just seeing things, or going out of her mind? Well, it was only to be expected. . . . But her fantasy proved to be sober truth. A convoy did in fact arrive with a large party of men, "politicals" and mostly Communists, who had been in solitary confinement at Verkhneuralsk. It was indeed our men—our very own! Thus the Social Revolutionary's prediction came true in part; but owing to the huge numbers involved in the operation, it was almost impossible to pick out one's own relatives. As I have already mentioned, Pava Samoylova, who found her brother, was the only lucky one among us.

Unhindered by the guards, we stood by the barbed-wire fence which separated our compound from the men's, and gazed spellbound at the long line of men who passed before us—silent, with bowed heads, plodding wearily in prison boots similar to ours. Their uniforms were also similar, but their trousers with the brown stripe were even more like convicts' garb than our skirts. Although one might have thought the men were stronger than we were, they seemed

somehow more defenseless and we all felt a maternal pity
for them. They stood up to pain so badly—this was every
woman's opinion—and they would not know how to mend
anything or be able to wash their clothes on the sly as we
could with our light things. . . . Above all, they were our
husbands and brothers, deprived of our care in this terrible
place. As someone expressed it, quoting from one of
Ehrenburg's early novels, "The poor dears have no one to
sew their buttons on for them."

Each face seemed to me to resemble my husband's; I was
so tense my head ached. All of us were straining to try to
find our loved ones. Suddenly one of the men at last
noticed us and cried out:

"Look, the women! Our women!"

What happened next was indescribable. It was as if some
strong electric current had flashed across the barbed wire.
It was clear at that moment how alike, deep down, all
human beings are. All the feelings that had been suppressed during two years of prison, all that each one of
us had borne solitarily in himself or herself, gushed to
the surface and mingled in a flood that seemed to be both
within us and around us. The men and women were shouting and reaching out to each other. Almost all were sobbing aloud.

"You poor loves, you poor darlings! Cheer up, be brave,
be strong!" Such were the words that were shouted both
ways across the wire.

After the first outburst we set about trying to find our
own people. This we did on a combined geographical and
Party basis, so that the roll call went like this:

"Leningrad regional committee!"

"Anyone from the Dnepropetrovsk regional committee
of the Komsomol?"

"Municipal committee of Ufa! Is your first secretary here?"

The next stage was the throwing of "presents" across the wire. The emotional tension on both sides needed an outlet in action: we each longed to give something, but we had no proper possessions to give. So one heard:

"Take my towel! It's not too badly torn!"

"Girls! Anybody want this pot? I made it from a prison mug I stole."

"Here, take this bread. You're so thin after the journey!"

There were also violent cases of love at first sight. As if by magic, these almost disembodied human beings recovered their sensibility, which had been dulled by such cruel sufferings. Tomorrow or the day after, they would be led off in different directions and never see one another again. But today they gazed feverishly into each other's eyes through the rusty barbed wire, and talked and talked. . . .

I have never in my life seen more sublimely unselfish love than that which was shown in those fleeting romances between strangers—perhaps because, in their case, love indeed was linked with death.

Every day the men would write us long letters—jointly and individually, in verse and prose, on greasy bits of paper and even on rags. They put all their insulted, long-pent-up manhood into the pure vibrant passion of these letters. For them we were a collective image of womanhood. They were numbed by pain and anguish at the thought that we, "their" women, had undergone the same bestial indignities as had been inflicted on them.

One of these letters began: "Dear ones—our wives, sisters, friends, loved ones! Tell us how we can take your pain upon ourselves!"

In spite of the huge population of both the men's and women's compounds, very few people met friends or acquaintances from their previous existence. Those who met others from the same part of the country considered themselves lucky. There was a funny little Tartar poet from Kazan, brought up in an orphanage, who rejoiced in the pen name "Republican Genius." I forgave him this in my joy at finding a fellow townsman, and we stood for hours on either side of the wire, repeating the names of Kazan families like a litany—a proof, for me, that all those poets, scholars, and Party workers were not figures in a dream, and that there were other people in the world besides jailors, camp guards, trusties, and goners.

This particular group of men met a tragic fate. Before they had time to recover from the train journey they were marched a distance of several miles toward the port where they were to embark for Kolyma. There was some hitch over the supply of bread that day, and the men were marched off on empty stomachs. After an hour or two in the broiling sun they began to collapse, and some of them died on the spot. The others sat on the ground and declared that they would go no farther unless they were given bread. . . . Organized protests of this kind were unusual among prisoners who had been Party members. The frightened guards went berserk, kicking the dead bodies after the manner of the good soldier Schweik's doctors ("Take this malingerer to the morgue!") and shooting several stragglers who "attempted to escape." However, the remainder were brought back to the transit camp for another week.

As usually happened after such an excess of severity, they began to feed the prisoners slightly better: the deadly "pies" became more frequent, the soup more substantial. The pies used to fly across the wire fence like tennis balls, as our kind friends kept trying to give them to us, but we refused them, saying we had plenty to eat. . . .

Allochka Tokareva fell madly in love with a boy from Kharkov, and spent whole nights talking across the wire. Her eyes shone fiercely, and all her sophistication was thrown to the winds. If necessary, she was prepared for a stand-up fight with Tamara herself, who, however, was contemptuously indifferent to these platonic, "literary" exchanges between the men's and women's areas. "Let them get on with it," she would say, "as long as the numbers come out right when they're counted. That's all that matters in the transit camp."

Our team that shifted stones from the quarry grew smaller day by day. More and more went down with diarrhea due to lack of vitamins, and they turned into so many ghosts. As a rule, the infirmary took only those who were clearly dying, and by no means all of them. The ordinary sufferers lay in a row on their bunks or on the ground, jumping up every minute to run to the latrine. Others, who had just come in from work and could hardly stand, brought the sick ones draughts of yellow, stinking water from the buckets which stood around, or, in despair, called the "sanitary orderly," who shoved tablets of salol into the victims' mouths with his dirty fingers.

The length of stay at the transit camp varied a good deal from one prisoner and one contingent to another: some were moved on within a few days, others remained for months. There were even trusties who conformed so successfully to the authorities' requirements that they were allowed to stay for years.

The roads out of the camp led in many directions. The Corrective Labor Camps Administration, North-East (CLCANE for short) was a mighty landlord, controlling vast areas. But as a rule the former inmates of political prisons went only to Kolyma. For some strange reason, the name of this place, which terrified everyone in the outside world, not only did not frighten us but was a source of hope. People would say:

"I hope we get started soon."

"At least they'll give us enough to eat there."

"The cold will be better than this scorching hell."

Such remarks betrayed the secret need of the human heart for hope, even of the flimsiest kind. We were also influenced by the tales about Kolyma which we heard from second offenders among the ordinary prisoners. Although they had been there as long ago as 1934–35, we drank in their stories greedily. Seasoned with a good measure of romancing and boasting, they depicted a sort of Soviet Klondike, where a man of initiative, even if he were a prisoner, need never go under, and where the most wretched invalid was quickly restored to health thanks to bountiful stocks of reindeer meat, red caviar, and cod-liver oil. Not to mention the gold, for which one could get tobacco and other stuff.

"Don't you worry, girls—in Kolyma there's space, food, and clothing for everyone!" said a yellow-haired young trusty whom we knew as "Vasek the Thief." Vasek's job in the camp was to check lists of prisoners: he always had lots of news, which he willingly shared. He was lyrical about Kolyma, where he had been twice—he was now in for a third time for having stolen money in Magadan.

"Feeling hot, eh?" he said sympathetically one day as

he met us sweltering on our way to the quarry. "Never mind, you'll soon be at Kolyma, it's nice and cool there." And he sang in his piercing voice:

Kolyma, Kolyma, you distant land
Where it's winter for twelve months in the year,
And summer for all the rest!

Like Gorky's character Luka,* Vasek never missed a chance to comfort his fellow creatures. He even had words of cheer for those suffering from night blindness, who listened to him eagerly.

"Never mind, girls, just wait till you get to Kolyma. You'll eat great hunks of walrus meat, it stands around in the camp in barrels just as water does here—solid vitamin A, just you ask the hospital orderly. Best possible thing for your eyes—after a bite or two you'll find the night blindness is gone, absolutely gone."

For older women, of whom there were a few in the camp and who suffered worse than the rest of us, he also painted a bright picture:

"Never you mind, Auntie, the great thing is to keep your courage up. You won't count as an old woman at Kolyma. You know the saying: forty per cent alcohol isn't vodka, a thousand miles isn't a road, a thousand rubles isn't money, seventy-five degrees isn't a frost, and no one's old at sixty! Don't you worry, Auntie, we'll get you married yet, you'll see!"

And although we all realized that Vasek's tales had to be, as the interrogators would have put it, "translated into 1937 terms," his description of the promised land was absorbed by many of us like a delicious poison. More and

* From the play *The Lower Depths.*

more often our feverish nights were interrupted not only
by groans and gnashing of teeth but by the cry "Why
don't they send us to Kolyma?" And Vasek, as he checked
his lists, winked at us more and more often and whispered
consolingly: "It won't be long now!"

4

· *The* S.S. Dzhurma

This was an old steamer that had seen better days. The
brasswork of the railings, the stairways, the captain's
megaphone, were alike dull and the brass parts had gone
green. She was used almost exclusively now for transport-
ing convicts, and there were ominous rumors about her,
for instance that if any of her passengers died they were
thrown to the sharks without even being put in a sack.

Before being taken on board we spent several hours
pitching and tossing in large wooden boats moored at the
quay, while the *Dzhurma*'s crew took their time preparing
for departure. We could see the sailors pushing heavy
string mops along the decks, while the captain and his mate
casually examined us through binoculars.

It was a gloomy, overcast day, with occasional shafts
of sunlight piercing the clouds. Dirty gray foam splashed
against the *Dzhurma*'s portholes. Even the physical atmosphere seemed full of anxiety and foreboding. None the less,

I felt curiosity as well—this was, after all, my very first sea voyage.

The time spent in the boats was uncomfortable and tiring. We were so crowded that my legs and feet grew numb; hunger and the sea air made me feel dizzy and sick. But the worst thing of all was the singing. Even now, twenty-five years later, I blush when I remember that amateur concert, although it was neither I nor my friends who had the idea of striking up those cheerful Komsomol songs. . . . Ira Mukhina was a ballerina, arrested under section 6 as the result of having had supper with some foreign admirers of her talents. The expanse of water before us reminded her of the Volga, and she sang:

Beauteous river
As broad as the ocean . . .

Others took up the refrain:

Free as our motherland . . .

"Shut up, stop that at once!" cried Tamara Varazashvili. "Where is your self-respect?"

"What do you expect from that crazy lot?" retorted Nina Gviniashvili with a look of disgust. The insulted singers raised their chant to a fortissimo. Anya Atabayeva, the former secretary of the Party district committee at Krasnodar, tried in vain to silence them in her deep voice, arguing that to sing of freedom in such circumstances might be interpreted as mockery and provocation. . . . The performance seemed to me more in the spirit of shameful self-abasement. I still shudder when I remember the captain and mate of the *Dzhurma* whispering to each other and passing the binoculars from hand to hand for a better look at our crazy amateur singers.

At last we got on board. Gangways, rickety ladders, steps up and down . . . If I remained on my feet it was only because there was no room to fall. We moved forward in a compact mass, and I was borne along like a single drop in a gray wave. I was ill, very ill. That morning I had had a high fever and acute diarrhea, which I concealed so as not to be left behind and parted from my friends. Now, as we boarded the ship, I kept fainting and I lived in a fragmentary, disjointed world.

Finally we found ourselves in the hold. The air was so thick and stifling one could have cut it with a knife. Packed in our hundreds so tightly that we could not breathe, we sat or lay on the dirty floor or on one another, spreading our legs to make room for the person in front of us. How we longed for the comfort of Car Number 7, where at least there had been bunks! . . . But surely the ship would sail soon? We heard the hull rubbing and squeaking against the pier, together with the noise of launches, tenders, and rowboats weaving around our vessel. The men were stowed in a part of the hold next to ours. It sounded as if the full complement of prisoners had been taken on board, and we waited for the ship to cast off.

But the worst was yet to come: our first meeting with real, hardened, female criminals among whom we were to live at Kolyma . . . It seemed as though there was no room left in the hold for as much as a kitten—but down through the hatchway poured another few hundred human beings, if that is the right name for those appalling creatures, the dregs of the criminal world: murderers, sadists, and experts at every kind of sexual perversion. I am still convinced that the proper place for people like this is not a prison or a camp, but a psychiatric clinic. When the mongrel horde surged down upon us, with their tattooed, half-naked bodies and grimacing, apelike faces, my first thought was that we had been abandoned to the mercy of a crowd of raving lunatics.

The fetid air reverberated to their shrieks, their fantastic obscenities, their caterwauling and peals of laughter. They pranced about and sang incessantly, even where there appeared to be no room to set foot. Without wasting any time, they set about terrorizing and bullying us "boobs" and "politicals"—delighted to find that the "enemies of the people" were creatures even more despised and outcast than they. . . . Within five minutes we had a thorough introduction to the law of the jungle. They seized our bits of bread, snatched the last rags out of our bundles, and pushed us out of our places. Some of us sobbed and panicked, others tried to reason with the girls or to call the guards. They might have saved their breath. Throughout the voyage we did not see a single representative of authority other than the sailor who brought a cartload of bread to the mouth of the hold and threw our "rations" down to us as one throws food to a cageful of wild beasts.

It was Anya Atabayeva who came to the rescue—a dark, sturdy woman of about thirty-five, with a deep, powerful voice and the arms of a stevedore. Letting fly with all her strength, she caught one of the women a resounding blow across the face. The girl went spinning to the floor, and for a moment there was a deep, astonished silence. Anya took advantage of this and, jumping onto a bale so as to be better heard, delivered in a voice of thunder such a stream of invective that the whole gang of them turned pale. The miserable creatures were as cowardly as they were depraved: Anya was the first of us to remember the saying "A lion among sheep, a sheep among lions." They were hypnotized by her personality and impressed by the form in which it showed itself.

"Who is she?" they asked one another, gazing at her with fear and admiration. From various corners of the hold members of our party called out: "The starosta!" This was enough for the newcomers: the fact that she could answer them in kind, and even, perhaps, have them sent to the punishment cells.

"Give back the bread and clothes!" Anya continued in a voice that struck terror; and they gave them back. They went on, of course, making a din, screaming and singing obscene songs, but their open aggression against us "politicals" had been stopped.

We had been sailing now for three days or so—it was hard to tell, for the days and nights were alike. When I opened my eyes, I saw rows of human faces with bloodshot eyes and pale, dirty cheeks. There was a revolting, sour smell. The sea was fairly calm, but the weaker among us could not help vomiting over their neighbors and the piles of dirty bundles. . . . For the first time in our sorrowful journey, which had now lasted almost three years, we had trouble with lice, for which we had to thank our new companions. Fat and white, they crawled about in full view, not bothering to hide in the seams of our clothes.

Luckily for us, our voyage was one of the *Dzhurma's* more uneventful ones. No fires, no gales, no flooding, no shooting at mutineers. When my comrade Julia, who had been too ill to sail with us, traveled on it a fortnight later, a fire broke out on board and some of the male criminals, who had seized the opportunity to try to break loose, were battened down into a corner of the hold. When they went on rioting, the ship's crew turned the hoses on them to keep them quiet and then forgot about them. As the fire was still burning, the water boiled, and for a long time afterward the ship was permeated by the sickening stench of boiled human flesh.

No such horrors took place on our voyage. We were even treated with a certain degree of humanity. Sometimes the hatch was left open, so that we caught glimpses of the sky, majestically still above us. And later, when more and more of us were suffering from diarrhea, we were even allowed to use the latrine on the lower deck. . . . Once, on the stairs which led to it, I fell and fainted. Coming to myself after a few seconds, I heard from somewhere above me the magical words: "Are you feeling very ill, comrade?" It was a cultured, man's voice, belonging to a doctor who was also being transported to Kolyma and who was in charge of the sick ward in the ship's hold. Was there really such a thing, I asked him, and how did they decide whom to put in it? Surely nobody in the hold was well? The reply was that you got there if you had a high temperature; and the doctor pronounced, at a guess, that mine was about 40°C. . . .

After a little further conversation it turned out that my new friend, whose name was Krivitsky, was not actually a doctor, but had been the deputy People's Commissar for aircraft construction. He had, however, lived abroad before the revolution as a political émigré and had studied medicine at Zurich. He had come to Kazan three years ago to open an aircraft factory. Did he remember Aksyonov, the chairman of the municipal soviet? Of course he did, and he had met his wife, a charming lady. "Do you mean to say that it was you?"

What he had said about the sick ward was true. It had

bunks in which all the sick lay squashed together in a row
—men and women, "politicals" and criminals, people with
diarrhea and syphilis, the living and the dead, whose
corpses could not be moved because they were out of
reach. In the corner stood an enormous bucket which
everyone, men and women alike, used in full view of one
another.

My temperature turned out to be 40.3, and on Krivitsky's authority I was squeezed in between a man and a
woman on the lower bunk. On my left was a burly, almost
naked criminal, who uttered fearful cries in his delirium:
he had a huge, fierce-looking eagle tattooed on his chest,
at about the level of my nose, and I kept thinking it would
peck me with its beak. . . . Groaning on my right was
Sofya Mezhlauk, the wife of Molotov's deputy. She
clutched my hand and repeated over and over again: "If
I die, please tell my daughter that I wasn't guilty of anything."

It was, I think, a Saturday evening, and there was a
party going on in the captain's cabin on deck. I could hear
the shuffle of dancing feet and a fox trot, incessantly repeated:

> Twilight falls softly,
> The stars shine bright,
> Kitty is dancing
> The fox trot all night,
> Tra-la-la! . . .

Once again I had the feeling that it was all happening
to me in a film. In a minute there would be a close-up of
the shuffling feet, and then they would show the bare
thighs of an old man sitting on the bucket: thin, trembling,
bluish like the legs of a plucked rooster. But no, it just

couldn't be done, it would be denounced as "naturalism." *

The drunken laughter got louder, the record started again. Was that really the only record they had?

I wanted to go to the lavatory, but I just couldn't here. They might be half dead, but they were still men. I would go on deck. . . . By a superhuman effort I raised myself on my elbows and slipped out from between my neighbors. It was nighttime. Krivitsky was sleeping in his own corner, on two stools placed together. How lucky he was to be a doctor! Just as well he was asleep: he would never have let me go up on deck; he had warned me that I might die if I tried to. . . . I crawled up the steep stairs, almost on all fours: it must have taken me a full hour. At last I could see the stars twinkling in the leaden-black sky, and the pennants of smoke from the steamer's funnels cleaving the darkness. Now I was on deck—I could see the water, and the ship's lights dancing above it. . . . Suddenly I realized I had lost my way. I knew the lavatory was close by, but could not remember how to get to it.

The same thing had happened to me after my spell in the punishment cell at Yaroslavl. It was a frightening sensation. A human being who has lost his sense of direction is no longer human. I fingered the walls like a blind person —the smoke seemed to be getting into my eyes, I could see hardly anything. But I must not die here at sea, to be thrown overboard to the sharks! O Lord, spare me till we reach Magadan, I beg You. I want to be buried in the ground, not at sea. I am a human being, and You Yourself said: Dust thou art, and unto dust shalt thou return. . . .

(Not long ago, in 1964, I read in a story by Saint-

---

* In Soviet literary doctrine, naturalism as opposed to realism is frowned upon.

Exupéry these words, spoken by a pilot lost in a storm over the Andes: "I swear that no animal could have endured what I did.")

I figured out afterward that this took place on the sixth day of our voyage. I was picked up by Krivitsky, who had wakened and noticed my absence: but I knew nothing of it at the time, for I did not recover consciousness till two days later, when the *Dzhurma* gave a cheerful blast on its horn on sighting, not far off beyond a ridge of hills, the outline of Nagayevo Bay. . . .

One by one, we invalids were carried ashore on stretchers and left on the beach in tidy rows. The dead were also stacked neatly so that they could be counted and the number of death certificates would tally. Lying on the pebbly shore, we watched our comrades being marched off toward the town, to the rigors of the collective bath and the disinfection chamber. We lay there till far into the night, and the guards began to curse their superiors, who seemed to have forgotten us "goners." It turned out later that there was a shortage of trucks, because several had been used that day to take parties of prisoners into the taiga.

Although it was August, the Sea of Okhotsk was of an implacable leaden hue. I looked around vainly on the landward side for a clean stretch of horizon—the purplish hills * fenced me in like prison walls. I did not know then that this was a feature of Kolyma: throughout the years I lived there, my eyes never once saw an open horizon. . . .

The guards, frozen and ill-humored, made a fire on the beach: its crimson glow was surmounted by black, resinous smoke.

---

* Literally *sopka*—small mounds of volcanic origin, which are a special feature of the Soviet Far East.

"Have you ever seen this place on the map?" The question came, in a casual tone of voice, from Maria Zacher, the Comintern German, who was lying next to me.

No, I did not know where it was on the map. I had indeed been monstrously ignorant in former days: I had never studied the map of Kolyma, nor heard of Kitty and her fox trot, nor known that people could be thrown to the sharks, as they were, without even being put in a sack. Now I knew all this. . . .

The sky began to glow with wonderful shades of mauve and lilac: my first Kolyma dawn was breaking. I felt, all of a sudden, a strange lightness and acceptance of my fate. True, this was a cruel and alien land. Neither my mother nor my sons would find the way to my grave. But all the same, it was part of mother earth—I had reached land, and need no longer fear the leaden, shark-infested waters of the Pacific Ocean.

5

• *No luck today, my lady Death!*

No, it was not a dream—I was actually sitting in a bath. I touched its dazzling white, slippery inner surface. What an amazing product of human genius! The hot water gave off a heady scent—it was a pine bath that the doctor had prescribed. Yes, the bath was real enough—but could this be my body, this emaciated thing that I could see through the water?

For the past two weeks I had been in that wonderland known as the Magadan camp infirmary, where I and others in like case were being treated, fed, and saved from death. And yet in the last three years I had grown accustomed to the thought that everyone I met who was not a fellow prisoner had only one aim: to torture and to kill.

The first few days I had spent here were a confused whirl of pain, blackouts of memory, and a dark abyss of unconsciousness. But one day, opening my eyes, I saw bending over me the face of an angel. Yes, it was just like one of those angels sitting on a cloud at the feet of the Sistine Madonna. However, the angel's blonde hair had a permanent wave and the soft chin was somewhat full, as befitted a woman approaching forty. My angel was appropriately named Dr. Angelina Klimenko, the wife of an NKVD investigator, who was in charge of the women's section of the Magadan infirmary.

"So you've come around," said my angelic doctor in flutelike tones. "You must eat as much as you can now. Never mind about the diarrhea."

The invitation to eat might have seemed in our circumstances a refinement of mockery, if she had not at the same time placed a substantial quantity of food on the table beside my bed. "Don't worry, you can eat it all," she added, moving off to the other patients.

I did not wait to be asked twice. I tore at the boiled chicken as greedily as, no doubt, my Neanderthal ancestors had devoured their hunks of roasted bison.

"What are you doing? Surely meat is bad for your diarrhea?" whispered Sofya Mezhlauk, who was my neighbor here also and who was convinced that the right thing for our condition was a starvation diet. But I trusted Angelina, who had said that the diarrhea was due to scurvy and that I should eat all I could. Most of all, I trusted my exhausted but young and incorrigibly healthy body, which was clamoring to be fed.

What induced Dr. Klimenko not only to keep me in the infirmary more than a month, so that I could recover fully from the journey, but also to bring me high-calorie food almost every day from her own home? Perhaps, as a doctor, she was fascinated by the process of resuscitating a half-dead person. Later on, she said to me several times: "When your party came off the ship, you seemed nearer to death than any of the others. I should never have expected that Zacher, Mezhlauk, and Antonova would die and that you would be the one to recover." Yes, no doubt she took a professional interest in my case, but that was not all. There were rumors that the angelic doctor had saved the lives of dozens of people by keeping them longer in the infirmary, having them excused from heavy work, or prescribing extra food for them. Moreover, I felt she had some personal sympathy for me. It was almost like a scene out of Dickens: an angel, in the midst of evildoers, saving my life. But there were times when I saw a look of sadness coming into her clear blue eyes; at such moments I suspected that it was not so much a Dickens as a Dostoyevsky situation, with Angelina seeking to expiate the crimes of a husband whom she loved. . . .

The days passed. Sofya Mezhlauk soon died, purely of starvation, having refused to obey the doctor's orders and eat. Although at death's door, she spoke with the self-assurance of a person used to being obeyed. "Nonsense, Doctor! I've been treated by Professor X in Oslo and Professor Y in Paris, and I know the only thing that will save me is dieting." Angelina, with her usual patience, did her best to explain that the diseases prevalent at Kolyma were not those commonly treated in Oslo and Paris, but Sofya only smiled condescendingly. . . . She died peacefully in her sleep.

The next to go was Maria Zacher. Before her death she suddenly forgot all her small stock of Russian words, even, for instance, the word for water. As the others in the ward did not understand German, and as I was now able to get out of bed, it happened that I was with her when she died. Her end was so "literary" that one would not have dared to invent it in a work of fiction. . . . Lying there, all skin and bone, she barely showed under the bedclothes. Her "Aryan" face was now sharply pointed: her nose and chin and the blue line of her lips stood out like Gothic lettering. Yet in this ghostly face her large brown eyes still glowed with the expression of her thoughts and suffering. To her last breath she was mentally active: a soldier of Thälmann's * army, anxious for the fate of the Communist movement. . . .

"Shall I be able to read Russian again? Why do you think I have forgotten the words?"

"Probably because your brain isn't getting enough nourishment. You'll remember them again all right."

A few minutes before she died, she began to recite some anti-fascist verses, I think by Erich Weinert, with the refrain *"Der Marxismus ist nicht tot."* Then she touched my hand with her bony, ice-cold one and said with her last breath: *"Aber wir sind tot."* †

Every day more patients died, both members of our

---

* Ernst Thälmann, 1886–1944, head of the German Communist Party.
† "Marxism isn't dead—but we are dead."

own and of previous groups. But this could not dispel the powerful sense of returning life which the rest of us felt. At all costs we must live—and each day brought something to be grateful for. First the diarrhea disappeared; then I put on four and a half pounds at one go, my cheeks became pink and my appetite still better. I found out that it was even possible to earn additional rations. A nurse called Sonya the Assyrian—one of the non-politicals—asked mysteriously whether I knew how to embroider. Summoning from the depths of memory what I had learned in my junior year at high school about needlework and crossstitch, I replied: "Yes, of course."

"Well, if you embroider this pattern on a cushion for me, I'll give you extra white bread, butter, and sugar." The pattern consisted of a bouquet of full-blown roses surrounded by the inscription in several colors: "Sweet dreams, Grisha, Sonya loves you."

Thus my days of convalescence were well employed. The roses were a success: Sonya was pleased, and every day she put on my bedside table something to eat. When I asked her where it came from, she gave a wheezy laugh and said: "Oh, you politicals are a simple lot! Just you lie in bed and get fat—what you don't know won't hurt you."

However, Grisha's dreams were not to remain sweet for long. One day Sonya, flashing her black Assyrian eyes and putting on my bedside a piece of Cracow sausage, said: "I want you to undo 'Grisha' and put in 'Vasek.'" Thus the vagaries of her love life assured me of another two days' work.

The real criminals were in a minority in our ward and behaved much better than on the *Dzhurma*. The atmosphere of the place put them in a sentimental mood, and in the evenings they would tell us their life stories, claiming that their father was a judge or a general, as the case might be, and telling tall tales of romance and crime; they revealed a certain poverty of imagination. They kept asking us either to "tell them the story of some book or other" or to recite Yesenin's poems.

One of the girls, the beautiful and brazen Tamarka, used to receive secret nocturnal visits from a gentleman friend who, after Ilf and Petrov's famous hero, I called Ostap Bender. During one of their meetings I happened to appear in the corridor.

"Watch your language," said Tamarka, looking at him fondly, "don't you see we've got a lady with us, an educated one too, one of the Article 58 lot?"

"Beg your pardon, Ma'am," said Ostap in an Odessa accent, baring a row of gold teeth as he smiled. "Beg your pardon. I have a great respect for learning. Believe it or not, I should really have been a corresponding member of the Academy of Sciences, but just now they haven't got any work to offer in my line."

"And what's your line?"

"Fireproof safes. I'm highly qualified, maybe you've heard of me?"

"He had a great name in Leningrad," said Tamarka proudly. . . .

Angelina began to treat me with injections, and I went on recovering by leaps and bounds.

"A lamb for the slaughter," said Liza Sheveleva sardonically. "Whom are you recovering for, may I ask? As soon as you get out of here, you'll go straight on to forced labor, and in a week you'll be the same sort of corpse that you were on board the *Dzhurma*. A fat lot of good Angelina's kindness'll be. It's just raising false hopes. . . ."

"That may be," put in Tamarka, "but we've a saying: 'You die today, and I die tomorrow.'"

"Neither of you is quite right," said Lucia Oganjanian

tactfully. "We needn't be either so selfish or so pessimistic. As Selvinsky said in one of his poems about a thief: 'No luck today, my lady Death! Till our next merry meeting!' And before that, perhaps his lordship Chance will take a hand. We've put off the evil day, and that's something to be grateful for."

When I returned from the infirmary to Hut Number 8 in the women's area of the camp, my first feeling was of shame. I could not bear to look at my comrades' livid faces, their frostbitten noses, cheeks, and fingers, and their hungry eyes, as they returned from work late on that November evening. Plump, rested, and well fed after my two months in hospital, I felt as if I had betrayed them.

After the separate beds, the clean floors, and the airy rooms, Hut Number 8 seemed like a wild animals' den. Everything was ramshackle; there were two rows of undivided bunks; it was freezing cold except around the huge iron stove in the middle which was always surrounded by stinking piles of jackets, sandals, and footcloths left there to dry.

"Been for a holiday?" asked Nadya Fedorovich spitefully. She was an oppositionist of many years' standing, arrested in 1933 and full of contempt for the "class of 1937."

The work to which I was assigned as of the following morning went by the imposing name of "land improvement." We set out before dawn and marched in ranks of five for about three miles, to the accompaniment of shouts from the guards and bad language from the common criminals who were included in our party as a punishment for some misdeed or other. In time we reached a bleak, open field, where our leader, another common criminal called Senka—a disgusting type who preyed on the other pris-

oners and made no bones about offering a pair of warm
breeches in return for an hour's "fun and games"—handed
out picks and iron spades, with which we attacked the
frozen soil of Kolyma until one in the afternoon.

I cannot remember, and perhaps I never knew, what
rational purpose this "improvement" was supposed to serve.
I only remember the ferocious wind, the forty-degree
frost, the appalling weight of the pick, and the wild, irregular thumping of one's heart. At one o'clock we were
marched back for dinner. More stumbling in and out of
snowdrifts, more shouts and threats from the guards whenever we fell out of line. Back in the camp, we received
our longed-for piece of bread and soup and were allowed
half an hour in which to huddle around the stove in the
hope of absorbing enough warmth to last us halfway back
to the field. After we had toiled again with our picks and
spades till late in the evening, Senka would come and sur-
vey what we had done and abuse us for not doing more.
How could the assignment ever be completed if we spoiled
women fulfilled only thirty per cent of the norm? . . .
Finally, a night's rest, full of nightmares, and the dreaded
banging of a hammer on an iron rail which was the signal
for a new day to begin.

It was the winter of 1939–40. One of us got hold of a
fairly recent number of *Pravda*, which caused a sensation
when we read it that evening before "lights out." It contained the full text, with respectful comment, of Hitler's
latest speech, and a two-page photograph of Molotov receiving von Ribbentrop.

"A charming family group," remarked Katya Rotmis-
trovskaya as she climbed onto the upper bunk. This was
careless of her: she had been warned often enough that
people had been put among us who listened closely to

what was said in the hut at night. Sure enough, six months later Katya was shot for "anti-Soviet agitation."

After ten days of "land improvement," the ulcer on my leg broke out again, and in an amazingly short time I was once more on my last legs, with no reason to feel ashamed to look my companions in the face. Angelina's efforts had indeed been wasted.

On Sundays we did not work, but washed and mended our ragged clothes and visited the other huts for people who had been sent here for lesser offenses and had received shorter sentences. These huts had a homely smell: tiny fish obtained from outside the camp were generally frying on the stove. Here and there on the bunks one saw blankets in check patterns and pillows with hemstitched linen covers. Most of the people here worked indoors, in the laundries, bathhouses, or infirmaries. They had healthy complexions and lively faces.

As a result of these visits I got to know the women of Hut 7, where amateur performances were given for the benefit of the camp authorities. A woman named Vengerova was a solo singer; there was also a choir, and former ballerinas took off their coarse jackets and sandals and put on gauze skirts to show off their art to the people in the front row. I happened to see one of these performances when I went there one Sunday. Thirty women who had been forcibly separated from their children and knew nothing of their fate were singing with great feeling as they pretended to rock a baby in their arms:

Sleep, my love, sleep, my little one,
The darkness does not frighten us,
No bogeyman shall come to hurt you.
Hushaby baby, hushaby baby.

They were congratulated by the officer in charge of culture and education for the excellence of their singing. One of the women in the hut, who occupied a trestle bed next to the stove, was the eighty-year-old Princess Urusova, a survivor from Imperial times. After the concert she remarked:

"When the ancient Babylonians led the Jews into captivity, they ordered them to play their harps; but the Jews hung them on the walls and said: 'We will work for you, but play we shall not.'" Shaking her almost bald head, she added: "There were no officers in charge of culture and education in those days. And altogether people were different then . . ."

In Hut 7 I used to hear various bits of camp rumor; the "terrorists" in Number 8 didn't have the heart for rumors.

"They say that a big party's to go off into the taiga shortly . . . to the state farm at Elgen . . . as a punishment. . . ."

"A lot of women are due in from Tomsk. Criminals' relatives—so far they've been kept there without any work to do, as if they were in prison. Now they'll have to work."

"The people from the prisons will be sent to the taiga, I expect. . . ."

One had to bear constantly in mind that however bad things were today, tomorrow they were apt to be worse. Each night, as one went to sleep, one could thank Fortune that one was still alive. "No luck today, my lady Death."

6

• *Light work*

Among the next party of women to arrive at the camp was a doctor from Kazan named Maria Nimtsevitskaya. Staggered by my miserable, scurvy-ridden condition, she offered me a pretty knitted jacket which, thanks to her profession, she had been able to keep.

We sat on the lower bunk in Hut 8, going through the names of friends and acquaintances. In almost every case they had been shot—got the standard ten years—vanished without trace . . . From time to time she stroked my hair, her face streaming with tears. . . . I fingered the jacket, quite spellbound by its shiny buttons and bold colored pattern. "I've got very thin, it'll be too big for me," I said, forgetting to wonder how it would go with the rest of my things: ragged sandals tied on with string, the gray prison skirt with brown stripes, the patched and shabby quilted jacket.

A rustling noise and the subdued voices of my companions announced the appearance of Verka, our team leader. Her sharp eyes immediately fastened on the jacket in my hand.

"Of course it's too big for you. What would you wear it for, anyway? Digging?"

She pressed the delicate wool in her practiced hand and watched it spring back into shape. "It's real wool all right. Let me try it on."

"Of course, Vera, do try it," said Maria loudly, squeezing my hand by way of warning. Verka, without further ado, shoved the jacket under her thick shawl and disappeared.

"Never mind, Genia," said my new friend earnestly, "that jacket may save your life. Of course some of the team leaders take things and give nothing in return, but I've heard that this Verka sometimes puts people on light work for as much as two weeks. And for you in your state, and after having been in the infirmary, it's so important not to have to go to that cursed digging. If you can get off it for a while, perhaps the worst of the freezing weather will be over."

Maria's prediction came true the very next morning. We had lined up at five o'clock as usual, blue with cold, and were waiting to be marched off. The sky was dark and overcast, without the faintest hint of approaching dawn. Falling into step with the others in my rank, I was moving off toward the gate when I became aware of Verka's attentive look. She was standing with a list in her hand, wearing her handsome sheepskin coat and thick shawl, and surrounded by a group of armed guards, at whom she smiled flirtatiously in the intervals of shouting at the team to get a move on.

On these occasions, Verka would not infrequently stop a particular group of five and order one of the ragged, sexless figures in it to fall out. This always brought our hearts into our mouths, as it could be either a good or a bad sign. It might mean that you were being picked out to be sent on the next taiga expedition, compared to which life at Magadan, digging and all, was heaven on earth; or it might mean assignment to an indoor job, where one could give one's swollen feet a rest for a few days and come into contact with people from the outside world, which in turn meant a chance to smuggle out letters, augment one's bread ration on the sly, and even scrounge a bowlful of soup now and then. . . .

"Fall out! On the left!" ordered Verka as I hobbled past her in my ragged sandals. . . . I could hardly believe my ears when, after the rest of the team had passed through the gate, she said to me casually: "You're to work in the guesthouse. Report to Anka Polozova."

The guesthouse—that fabulous corner of the free world, that Arcadia to which non-politicals were sometimes sent as servants but which we Trotskyists were not normally allowed to enter; where, after cleaning the floors, one might arrange to do residents' laundry and get in return white bread and even sugar. . . . Evidently Verka, unlike many of her kind, was a high-principled taker of bribes, who gave value for benefits received.

At this time the guesthouse was a large gray barrack, two rooms of which were occupied by minor officials who were living there with their families while waiting for their quarters to be built. Apart from them, the place was inhabited by some of the old-timers of Kolyma: drink-sodden mining clerks, criminals on parole between sentences, and even a few adventurers who had found it a convenient place to settle with the aid of forged papers.

The rooms were full to bursting, and so were the corridors, which were lined partly with mattresses and partly with camp beds sleeping two apiece. The corridor population consisted mostly of geologists who had been arrested in 1937 and set free in the "liberal spring" of 1939, and were now waiting for the spring and the melting of the ice so they could return to the mainland.

"Now mind, girls," Anka Polozova instructed her team of five criminals and myself, "up to three in the afternoon you're on duty cleaning the place, but from three to lights

out you can do what you like—only if you get caught it's
your own affair, don't expect me to help you out." Anka
herself had the convenient rank of a "socially harmful
element," midway between a "political" and a common
criminal, and hence eligible for the task of controlling a
team of cleaning women. "And now," she went on, "I
must go and do my accounts."

"That's right," muttered one of my new colleagues,
Maruska by name. "You go off and do your accounts. The
poor fellow's waited long enough, his eyes are nearly
popping out of his head."

Indeed, the assistant manager of the establishment, a
burly Caucasian named Ashotka, with goggle-eyes and a
pointed chin, with whom Anka had been "doing accounts"
for the past month, stood waiting for her in the doorway
of his room.

"I'm coming, honey-bun," said Anka with unexpected
tenderness. "Now listen, you lot: this girl here's new
today, she's a political and has had a rough time of it, so
don't play any tricks with her. What's your name? Genia?
Right. Go with Maruska here, she'll show you the ropes.
Now I must go and do my accounts. I'm coming, sweetie-
pie!"

"Some sweetie-pie," muttered Maruska, rolling her
dreamy blue eyes. "I'd sooner have the job of burying
him than feeding him—he eats like a boa constrictor, and
at our expense too!"

That evening I saw in operation the unwritten law
whereby half the money the girls earned on the side went
to feed the Caucasian boa constrictor and his lady-love.

Our work consisted in washing the much-trampled, unpainted floors. I took my bucket and cloth and queued up
in the boiler room, whose aged custodian (also a prisoner)

carefully meted out to each of us half a bucketful of hot water, leaving us to fill up with snow. After looking sideways at me once or twice from under his shaggy brows, he correctly guessed the article under which I was convicted and the length of my sentence.

"You were in solitary, weren't you? Tha-at's it. . . . Better take off those hemp sandals, the water'll spoil them. Here, you girls, surely you can give her something better to wear?"

"Don't worry, Grandpa, of course we can. Here, Genia, take those things off. Here's a pair of galoshes for you." A kindly woman called Elvirka, who was tattooed from head to foot, kicked off the down-at-heel pair of men's galoshes in which she had entered the boiler room.

"It's most kind of you, Elvirka, but how will you manage yourself?"

"Christ, isn't she polite? You politicals sure are a funny lot. 'May I ask your ladyship in for a cup of hot water and a chat about the latest books? What a delightful day it is, to be sure!' . . . Don't worry about me, dearie, I'll get another pair from some sucker. I shan't go barefoot." With an apelike movement Elvirka started scratching the sole of her right foot, which was tattooed with the inscription: "Don't forget your poor old mother."

As I was walking out behind the others, the boilerman stopped me, saying:

"Pleased to meet you. As you're a political, I just wonder how you ever got sent here. Medical certificate? I'm from Leningrad, I'm in for anti-Soviet agitation too, Article 58, section 10. I got here on a medical certificate too. I had an operation during the civil war. Now my guts are all coming apart. After the journey here the stitches came loose, so they took pity on me and gave me this soft job—anyway, I'm over sixty. But never mind about me—

what I wanted to say was, you'd better watch out here, this is no place for a young girl like you."

"Thanks for the warning, but actually I'm over thirty— I look like a girl because I'm so thin."

"Well, that's young enough, and this place isn't right for you. I've seen what goes on here. Don't ever go into any of the rooms. That Ashotka's the worst of the lot. If you want to earn a bit on the side, try the women— there are two married couples in the place. When you've done the floors, come and see me and I'll take you to the Solod woman—she was asking me yesterday about getting her laundry done. She's as mean as hell, but you'll get a bite extra to eat. And you can safely work for the fellows who sleep in the corridor, they're more our kind, reha- bilitated and all that. They're in pretty poor shape, poor devils, they've had two or three years in camp, but they'd share their last crust with you, and their things need wash- ing and mending badly."

My companions got through their assigned work two hours before I did. Then, dressed in baggy trousers or dresses of fantastic pattern, with scarves tied low over their foreheads in a special knot, they began to chase all through the building, which rang with their shouts, laugh- ter, and obscenities. They meant no harm by these: their attitude was peaceable and even friendly, but they could not express any thought without a liberal sprinkling of four-letter words.

"A pretty low form of life they are," said the boilerman tolerantly, as I went to fill my bucket once again. "Those are all the words they know. Some of them aren't bad girls, though, if they're taken in hand early enough. Yes, back in civilization we had no idea of all the riffraff there is in the country."

Washing the floors was not a difficult job, though I felt

dizzy from hunger and from bending down. The work was a joy in comparison with outdoors: the heavy iron pick with which one desperately attacked the hard-frozen ground, the frost savagely piercing one's threadbare jacket. Here at least I was indoors and warm, and the hot water helped to heal my swollen hands. None the less, I felt like crying when the guests rushing along the corridors left dirty footmarks in places I had just washed.

"Hey, your ladyship, what on earth are you up to?" Elvirka, now heavily made up and dressed in a raspberry-colored housecoat with a flower pattern, surveyed my work with unconcealed astonishment. "You'd think you'd gone to stay with your mother-in-law and wanted to show her what a good housewife you are!"

"Well, instead of shouting at her, why don't you show her? You know she's a political, and they're all pretty well done in." This was Maruska, her blue eyes contrasting strangely with her hoarse drunkard's voice. Pulling me by the sleeve, she said: "Look here, Genia, to begin with, you wouldn't get this place really clean in a month of Sundays. Secondly, why work for the boss when you ought to be making a bit for yourself? Thirdly—look, this is the way to do it." With a deft movement she emptied the bucketful of water onto the floor and spread it about with broad sweeps of the cloth. "Hurry up, so that Ashotka can see it's been washed, and then come and have tea in the boiler room—I've scrounged a bit of white bread."

It was a blissful sensation to sit by the warm boiler drinking near-boiling water from a mug provided by the old man, and nibbling from time to time at a piece of sugar or bread from Maruska's ration. . . .

The Solod woman was not one to mince her words.

"This one? Why, she can hardly stand. At death's door

if you ask me. How could she possibly get through that pile of laundry? It hasn't been touched for a month!"

"If people are made to do hard labor on an empty stomach, they get thin," said the boilerman sententiously. "But she's a willing worker, and she won't charge you much."

Almost lovingly I set about sorting the Solods' laundry. This was a turning point, a moment of triumph in my existence as a prisoner and camp inmate. For the first time in three years I was about to earn my bread independently and on my own initiative. Moreover, the work itself was of a sensible kind—to provide a clean change of clothes for the scruffy children swarming about the disorderly room amidst piles of unwashed dishes.

"Are you sure you haven't got some infection or other?" asked their mother. "You look very thin to me. . . ."

"Yes, quite sure. Scurvy isn't infectious. It's because I don't get enough to eat."

"All right, then. I'm going to the shop now, and then I'll be back for dinner." She whispered for some time to her elder child, a boy of about ten, casting sidelong glances at me. Soon after she had gone he slipped into the corridor, saying over his shoulder to his six-year-old sister: "I'm tired of seeing she doesn't steal anything, you watch for a bit."

Julia, my companion at Yaroslavl, had been right when she said jokingly that if I was given a quarter of a pound of solid food I immediately put on two pounds or more. After I had been a week at the guesthouse I was already unrecognizable.

"Heavens, how the girl's filled out at my expense!" said the Solod woman almost good-naturedly, as she gave me a second helping of *kasha* with plenty of fat in it. Within a week I had got through most of the work in the place and she was the more appreciative when she found that none of her property was missing.

"You're quite good-looking after all, with those big eyes of yours. I don't see you being a laundress and cleaning woman for ever—there's a shortage of women here at Magadan, but the ones in this hotel are real harpies."

I tried to make her understand that I was not "that sort."

"Very well, then," she said approvingly, "you put on a bit more weight and we'll find you a decent husband. Why, you even get clerks from the mines here—white bread, butter, and sugar, and money in the bargain. . . ."

When I returned to Hut 8 in the evenings, I used to amuse my companions with descriptions of the hotel characters. I myself did not dream of taking seriously the Solod woman's plans for marrying me off to a "well-to-do" mining clerk.

One day, however, as I was mopping the corridor (speedily, by Maruska's method, so as to have more time for the Solod family), I suddenly felt a resounding slap on the behind and heard a hoarse, alcoholic voice saying: "Well, what about it? I'll give you a hundred rubles."

Up till then, the question of prostitution had come my way only as a social problem (in connection with unemployment in the U.S.A.) or as the subject of a movie with Alisa Koonen in the foreground under a dim, swinging lantern. Even in the most lurid nocturnal visions of Butyrki and Yaroslavl I had not imagined that I myself could be the object of such words and gestures. I was so shaken that I forgot the boilerman's precise instructions on how to deal with this kind of emergency ("Slap him across the face with your floor cloth and tell him to go to hell"), but instead I produced from some depths of my subconscious the words: "You scoundrel! How dare you!"

A grin spread over the brown, peeling, frost-nipped cheeks of my would-be customer. He tilted his cap on one side and continued:

"My! What a pair of eyes! You're a beauty, you are. Are you Article 58? All right, I'll go to two hundred."

His blue, frostbitten hands with their crooked fingers reached for me again.

"Keep off," I screamed, seizing my bucket, "or I'll pour this over you."

Suddenly I was aware of an arm in a leather sleeve and a hand which seized my cave man by the nape of the neck like a kitten, after which a well-planted kick (the foot was shod in good-quality felt boots) sent him flying to the end of the corridor, from where he filled the air with a volley of curses.

My rescuer was Rudolf Kruminsh, one of the corridor lodgers who were waiting for sea transport home after two years of detention. This incident marked the beginning of my friendship with him and his fellows. From now on I would hurry through my work for the Solods in order to have at least an hour with them before it was time for me to return to camp. As quickly as I could I washed their underwear, sewed on their buttons, and washed their cups and bowls.

The corridor was an oasis—a place of friendly, human faces, conversations about the mysteries that perplexed us all, and absolute trust. None of my new friends hesitated to take the risk of posting letters for me outside the camp.

"Genia," said the swarthy geologist Tsekhanovsky one day (he had a permanent cough as a result of ill-treatment during interrogation), "you shouldn't take so much trouble about sewing the buttons on that leather coat of Rudolf's.

It's a waste of time, because he cuts them off again every night."

Rudolf had been transferred at an early stage to administrative work and was much better dressed than the others. I had been wondering why his buttons kept coming off so frequently, not realizing that it was a pretext on my rescuer's part to reward me each time with candy and bits of sugar. A blush spread over his rugged face as he growled at Tsekhanovsky: "You're a real bastard."

Now I would jump out of bed in the morning without the usual feeling that life was not worth living. I waited with impatience for our squad to be marched out before dawn, and felt relieved when the camp gates closed behind us. I kept close to Elvirka and Maruska as we hastened toward the guesthouse along the dark-gray, foggy, frozen streets of Magadan. In that menagerie where petty "officials" from the camps of Kolyma, clerks, and prostitutes were thieving, drinking, and whoring, I looked forward to meeting the kindly faces of comrades who had escaped from the grasp of the monster that was destroying me, and whose unselfish care enabled me to be fed and warmed, body and soul.

I kept chasing away the sneaking fear that this blissful life might come to an early end; and it was a heavy blow when, as we were marching out past the armed guards one bitingly cold December morning, I heard Verka's voice calling:

"Fall out! On the left!"

So it was over. Well, a month in the guesthouse was pretty good exchange for a woolen jacket with shiny buttons. . . .

"Back to the hut for today! Tomorrow you'll go on field work as usual."

I lay till evening motionless in the empty hut. The sharp pain in my heart was not due so much to the thought of the rusty pick and the stupefying cold as to the knowledge that I should never see my corridor friends again, never hear Tsekhanovsky's jokes punctuated by coughing, never sew on the buttons which Rudolf was at such pains to cut off again. . . .

That evening, just before the working party returned, the door of the hut opened, and in the swirl of fog I made out the figure of Anka Polozova in her elegant felt boots. She looked around conspiratorially and said:

"Sh-h! Listen—you're not to worry. Your pals over there haven't forgotten you. Here's something from the one in the leather coat. Bread, butter, sugar—and here's some money. And now look here . . ." She pulled a heap of crumpled notes out of the pocket of her smart quilted jacket. "This is the main thing—it's for Verka, so that she won't send you on outdoor work again. Of course you won't get back to the guesthouse: she got hell from the camp administration for sending you there in the first place, but she'll think of something."

"Who's the money from?"

"From your friends. They argued for an hour or two about whether it's what they called ethical or not. But in the end they passed the hat around and they told me to tell you not to refuse it. They said it's all right for someone in your place to do what you can to save yourself; if you don't you'll croak."

So when the all-powerful Verka next told me to fall out, it was for "kitchen duties" in the men's zone. After she had been bawled out for allowing so dangerous a character to work for a whole month not under guard, she now arranged for me to be employed as a dishwasher in the male prisoners' canteen, where the work was light and I had a roof over my head. I did not have a chance to worry about the ethics of bribery in camp conditions, since Anka took it on herself to pass the money to Verka.

The job I had been given was one of several, ranging from dishwashing to sweeping snow off the Magadan streets, which were normally reserved for the living skeletons who were taken off hard labor because they were too sick, the so-called "goners."

The manager of the canteen was a Crimean Tartar called Ahmet. His handsome face, with its eyes like olives, was a study in cunning and shrewdness. His speech, gestures, and habits were those of the resourceful servant of picaresque comedy. Before his arrest he had already been a cook or, as he called it, a "chef cook." All day long he raced like mad around his kitchen, singing to the accompaniment of the clatter of knives and forks. By shamelessly starving the half-dead prisoners, he managed to feed himself and his retinue quite well out of camp rations.

The problem of women was an acute one for the arrogant, well-fed trusties recruited from among the "embezzlers" who held positions of authority in the men's zone. Although the two or three female dishwashers and kitchen helpers were fed on large quantities of stolen meat, they still could not keep pace with the heavy demands made on them. The contrast between the wretched goners who were barely able to walk and their rapacious masters made this work extremely distasteful to me. On the first day I was constantly close to tears; I realized that I was surrounded by wild beasts and should be lucky if I could keep myself from going under for more than a few days.

My appearance—a woman, and a "political" at that—caused a sensation. I arrived at 6 A.M. and sat, anxious and depressed, waiting for Ahmet, lord and master of the kitchen, to emerge from his private cubicle. The canteen and kitchen were filled with the rancid smell of oatmeal soup and green cabbage leaves. I sat there like a person condemned, while an unsavory pack of trusties debated with relish which of them would have first go at me.

Clearly I would have to get out of this den of wolves even if it meant going back to "land improvement." I looked around in despair for a savior of Rudolf's sort, but saw none—they were all criminals of the lowest type, the aristocracy of the camp world.

"Beat it, everybody!" shouted Ahmet in a high but deafening voice. "She was sent to the canteen, wasn't she? And who runs the canteen, I'd like to know! Get out of here!"

He looked me over and, humming an underworld song, he did a little dance and laid at my feet the fabulous gift of a bowlful of dumplings. According to regulations these were supposed to be given to goners as a reward for good work, but in practice they were usually devoured by the trusties.

The only weapon in self-defense against animals like these was cunning. I decided to surprise Ahmet by displaying my knowledge of the Tartar language. With a great effort of memory I managed to string together a few phrases to the effect that I was from Kazan, that I was practically a Tartar myself, that I was worn out from prison, that he should treat me like a sister and not leave me to the tender mercy of the others.

It was a long time since Ahmet had heard the sounds of his mother tongue, and an almost human expression flickered in his olive eyes. A real Moslem woman—well, I'm damned! What a stroke of luck! Worn out by prison, you say? We'll soon fatten you up. Ahmet can wait a week—just you work away and eat all you can, nobody'll touch you. Eat plenty of dumplings, that's an order, we've no use for thin ones here!

A week—well, that was some respite. Perhaps by then the stabbing pains in my chest would have stopped, and I could go back to the heavy work of land improvement.

"Eat away, eat your bellyful," said Ahmet, pushing a hunk of boiled meat onto my plate. "Don't bust yourself as far as work's concerned, you've got a mate who's as strong as a horse."

Turning my head, I saw a middle-aged man busying himself with quick, automatic movements about the sink, which was divided into two parts. His cultured face was covered with a dark stubble; his lips were tightly closed, and a cap was drawn down over his forehead. The plates and dishes were pushed through an opening in the wall and clattered into the sink; the remnants of soup were washed off them on one side, and they were rinsed in the other, dried, and pushed back to be refilled at a central counter. We had, of course, no running water, and every ten minutes or so we had to fill two buckets with hot water from the boiler room across the yard.

I at once noticed the thorough, efficient manner in which my gloomy companion was working: the dishes passed through his hands in an unending series, as if on a conveyor belt. He looked well fed and was clearly not a goner like myself—how had he managed to land a woman's job like this and not be sent to the mines?

"He's deaf," said Ahmet, noticing my glance. "Deaf as a post, you could fire a gun and he wouldn't turn a hair. Been through one medical commission after another. He's a Volga German, he can do as much work as two men.

You do the rinsing and leave the water carrying to him. Just you get up your strength, then we'll have another talk, you and I. No one'll interfere with you." He winked significantly.

"Is he dumb, too?"

"No, only deaf—he jabbers to himself in his own lingo."

I listened to the deaf man's mumbling and distinctly caught the word *verflucht*, applied to Ahmet. . . . I set about the job of rinsing, which was not so easy as it sounded. The tin plates sped around at an alarming rate, and I kept repeating the same monotonous movement of my arm. After two hours my neck and shoulders were numb. I naturally did not want to take advantage of the German's capacity for work as Ahmet had suggested, and tried to make him let me fetch water; but he firmly took the buckets out of my hands, mumbling as he did so: "The damned goat, plaguing the life out of the women," and casting angry glances at the "chef cook."

The days passed by in this manner, and at night in my sleep I saw the greasy plates come at me. The inmates of the camp worked and ate at different times, so that the canteen was always functioning. At peak hours one could not take one's eye off the sink for a second, let alone think of stretching one's weary limbs for a moment. The sour, sickening smell of bad soup had now got into my hands, dress, and quilted jacket. Owing to the hot water I was constantly bathed in sweat, and whenever the door behind my back opened an icy blast swept through it. I was plagued by a cough which prevented my sleeping at night.

But when Ahmet from time to time took pity on me and changed my job to that of collecting mess tins from the canteen, things were even worse. The place and the people there were alike horrifying to look at. There were not

enough seats for everybody, and many of the men ate
standing around the big iron stove. The hands in which
they held their food trembled. The stink of the hempen
sandals, drying in the warm air, was more overpowering
even than that of the foul soup. How they trembled, those
bony hands, black with frostbite, as they clutched the mess
tins! The place rang with noise: cursing, hawking, loud
coughing, the rattle of spoons. Worst of all were the
prisoners' jokes:

"Here's to us, and let's hope it's not the last!" they
would say before downing a glass of the anti-scorbutic
plant extract which was doled out to them.

Ahmet was a stickler for hygiene and neatness. The
warped, ice-cold windows of the canteen were hung with
patterned curtains, and at one time he had even seen to it
that a washbasin and towel were provided, together with
a notice in artistic lettering announcing that washing one's
hands before a meal was the way to avoid scurvy. The
notice was still there, but for a long time, now, no water
had been put in the basin. The goners were past the stage
where their hands could have been washed clean.

The trusties ate in a special corner in the kitchen from
which tantalizing odors drifted over to us of real meat
broth and excellent dumplings fried in sunflower-seed oil.
On my first day Ahmet wanted to seat me with them, but
I begged him with tears in my eyes to let me stay with the
deaf man. "I'm afraid of them," I said with perfect sincerity. Ahmet took my fear of the ape-men as a sign of
the bashfulness of a Moslem woman, and explained solemnly to one or two of them that the women of Kazan
were not trollops, whatever might be the case in Moscow.

Of course, if I had been strictly honest and high-principled I would not have eaten the dumplings, made with

stolen flour, which were supposed to "strengthen" the soup. But I was too hungry to achieve such heights of virtue. So, salving my conscience with the rather despicable sophistry that if I didn't eat them the goners wouldn't get them either, I carried the dumplings off to our scullery and placed them on the overturned box, covered with newspaper, on which my deaf companion had already placed a few hunks of bread. Then we sat on opposite sides of the box, on stools turned over sideways, and ate soup out of a single bowl, as this was the easiest way to divide fairly the small pieces of reindeer meat that occasionally turned up in it. Any sense of squeamishness, or the rational fear of contagion as a result of sharing a dish with a stranger, had disappeared from our minds long ago. In any case, I felt instinctively that the deaf man was clean in all senses of the word. From the second day on, a tacit understanding had grown up between us. I was amused and touched by his way of treating me like a lady in this jungle world. He stood up when I did, let me go through doors first, and helped me on with my prison coat as though it were sealskin. . . . He continued to talk to himself a great deal, of course in German. Being used to the fact that everyone regarded his speech as a senseless and unintelligible mumbo jumbo, he did not hesitate to express his thoughts aloud. Listening to him, I soon discovered that he was a devout Catholic from a prosperous farming background. He communicated with me by means of gestures and mimicry, having no idea that I knew German. This made me feel awkward, as if I were eavesdropping on his secrets. One day I tore off a piece of newspaper and wrote on the edge in German: "I can understand all you say, please remember that."

Helmut (he told me his name that day) was much

moved. He looked at me for a long time with his moist eyes, then kissed my hand, swollen as it was from dishwashing and reeking of soup, and said he was certain the *gnädige Frau* would never give him away: he could see it in my face.

Not long after that, an event took place that intensified Helmut's feelings toward me. It was a dramatic event for me too; for a time it restored the sharpness of my response to our life and all its horrors, a response which had been blunted by camp life with its daily struggle for existence.

One morning a party arrived at the camp consisting of men who had been worked to complete exhaustion in the mines, human slag that was now of no further use there. On the march back, numbers of them had died like—I was going to say "like flies," but at Kolyma it was truer to say that flies died like people. The survivors were sorted out at Magadan, where a few remained; the majority, however, were assigned to "light work" at such places as the Taskan food-processing plant. Here, before being released into a better world, they spent twelve hours a day in a fifty-degree frost cutting branches of a plant used for producing the anti-scorbutic extract made in the factory.

As usual when such parties arrived, there was a great emergency in our kitchen and canteen, preparing extra quantities of soup and white bread and washing piles of extra bowls. As I stood bent over the sink, a man poked his head through the opening; he wore a cap with a dirty towel wound around it.

"Which of you's from Kazan?" he asked in a hoarse voice. I trembled at the wild thoughts that went through my head. Could my husband be among these dying men, or was it a message from one of my friends, and if so, who?

"One of the fellows here is from Kazan—he's on his

last legs, he won't see the night through. He heard there
was a Kazan woman working here, and he sent me to ask
if there was a chance of his getting a piece of bread for
his last meal. Could you spare some for him? You're so
lucky to be where the food is."

His voice trembled from a mixture of acute envy and
humble adoration of anyone who occupied such an exalted
position in life—"where the food is"!

"He promised me half," the man went on, rubbing his
forehead and cheeks with a filthy sleeve as they started to
sweat because of the heat from the sink.

"Here you are," I said, handing him a ration. "Give him
my good wishes. Wait a minute—who is it, anyway?
What's his name?"

"He's a Major Yelshin. He worked in the Kazan
NKVD."

The bread in my hand trembled and dropped to the
floor. Major Yelshin! I saw, as if in close-up, the comfortable office with the big window looking out on the Black
Lake. I heard the velvety tones of the Major's voice:
"Make a clean breast of it. . . . You're just being romantic . . . taken in by those filthy subversives. . . ." It was
he who had decided that my crime fell under the article
about terrorism that carried the death sentence. It was he
who had put me in the fearful category of prisoners sent to
solitary confinement. I understood that it was not in his
power to have me released if he didn't himself want to fall
foul of the "wheel of history," but all the same, it was up
to him whether I got five years or ten: he didn't have to
brand me as a terrorist, he could have kept it down to
"anti-Soviet agitation," which gave one a better chance
to survive. And those sandwiches—could I forget those
French rolls with slices of tender, pink, succulent ham

that made one's mouth water; how he had put the plate before me, a hungry prisoner from the cellars, and tempted me with the words: "Just sign, and you can eat as much as you like."

"What's the matter? Did you know him? He doesn't seem to have been such a bad fellow. A lot of other NKVD types have been sent here to the mines, but no one seems to have had it in for this one. Anyway, does it matter now? He'll certainly be dead by the night, I can tell only too well. Once their teeth get long and start sticking out of their mouths, it's all over."

A shadow of fear passed across the messenger's sunken eyes—was the half-ration going to slip from his grasp at the last moment?

The teeth—that detail was all that was needed to overcome my hesitation. Sticking out of their sockets with the dry gums drawn back—just as I had seen Tanya's in the transit camp. . . .

"Here's the bread. Give it to him. . . . Wait a minute, though—you're to tell him it's from me. Remember my name and repeat it to him. . . ."

Suddenly my legs gave way under me. I sat down on the box we used as a dining table.

"*Was ist los?*" asked Helmut anxiously, offering me paper and pencil for my reply. I wrote:

"The man asking for bread was my investigator."

"*Ach so! . . .*"

During the next few days I suffered terribly, not knowing whether he had died—Yelshin, the elegant major whose task it had been to offer the carrot while others plied the stick. What made me suffer was my own behavior. How could I have been so petty as to insist on his knowing my name, to poison the last mouthful of bread that he would eat in his life? How despicable of me! Surely in this inferno we were quits—our accounts were closed once and for all by his death, by a death like this!

Yet, while I was tormented by these thoughts, Helmut was strangely exalted by the episode of the bread ration. During our work he whispered to me in German: "Your life will be saved, do you hear me? You will come out of here alive, because you gave bread to your enemy. I am your friend for ever—I would give my life for you."

Unfortunately, he was very soon to have a chance of showing that he had spoken these words in earnest. . . . It was nearly the end of the week which Ahmet had given me to "get up my strength." More and more often, I felt his greedy eyes on me. When, one morning, he turned up strutting like a peacock and offered me a large woolen scarf (it was easy for trusties to steal such things from new arrivals in the disinfection centers), I realized that my breathing space was over and that I must resign myself to going back to outdoor work.

"No, thank you, I really don't need a scarf—my camp one is warm enough."

Ahmet's mouth shut tightly, like a trap. "All right, you're a lady, I know that, that's why I waited and fed you up. A lady today, a lady tomorrow—how long can you go on being a lady, eh?"

He walked away in disgust. An hour later he summoned me to his quarters to get a new floorcloth, for which I had asked him some time ago. It was a handy pretext. I felt afraid, but reassured myself: he wouldn't dare with so many people within earshot, I'd scream if he . . . Nevertheless I scribbled a note for Helmut: "Ahmet's called me to his quarters, please keep an eye on me!" He nodded reassuringly and his deep-set eyes lit up fanatically.

A small red bulb was fixed to the ceiling of Ahmet's room. He himself was sprawling on a heap of sacks with the air of a pasha taking his ease.

"If you don't want the scarf, look at this!" In his hand, tinkling and sparkling, was a long necklace of cheap glass beads. Evidently he felt that this gave adequate expression to the ardor of his chef-cook's heart, and he must have reckoned that the prize was as good as his. My refusal to accept his little present aroused the cave man in him. I rushed to the door, but found it locked. I screamed as his mouth, like a trap closing, and his glittering eyes drew nearer. Suddenly the flimsy door began to shake and creak violently; there was a final heave, and I saw Helmut lying on the floor with the door under him, as though hurled forward by the anger which blazed in his face. He looked like a wounded gladiator or a medieval huntsman who had just killed a wild boar.

After a moment's silence there was an explosion of Tartar and German oaths. Before long Ahmet found his Russian again and shouted:

"I'll show you, you pair of bastards! So the deaf man's better than Ahmet, is he? I'll have you both thrown out of here and sent to a place you'll never come back from, by God I will!"

But Ahmet's revenge had to be put off for a few hours, as just then we were warned of the return of a huge new batch of prisoners from the mines.

"Hurry up!" said their starosta who brought this news. "They're to be fed immediately, they are dying on their feet and you'll have to answer for them. What? Taking these two off the job? You must be crazy. Here, get moving right away, they must all be fed in half an hour!"

Ahmet got moving. "The deaf man can wash dishes by

himself," he ordered, "you've played enough games together. And you," he said to me, "go in there and hand
out the bowls."

So I stood and dished out the food, methodically dipping
my ladle into the caldron of soup and holding out a mess
tin to each of the weird creatures that passed before me,
muffled in rags and bits of sacking, with black, frostbitten
cheeks and noses covered with running sores, with bleeding, toothless gums. Had they issued from primeval night,
or from the sick fantasy of a Goya? I was paralyzed by
horror, but I went on mixing and stirring the soup in the
caldron so that what I ladled out to them might be as thick
and nourishing as possible. . . . Still they came. There
was no end to the black procession. With their stiff fingers
they took the bowl, stood it on the edge of the long, plank
table and ate. They partook of the soup as though of a
sacrament that held the secret of the preservation of
life. . . .

One of them leaned through the hatch and said: "Hey,
you, make it as hot as you can! Something to warm our
guts!"

"It's hot, comrade, I hope you'll enjoy it," I replied,
weeping. I heard him shout:

"Well, of all things, it's a woman! Here, Mitka, come
and look, a real woman! My God, it's three years since a
woman served me soup!"

This was someone very different from Ahmet the chef
cook: a simple Russian peasant, the father of a family, who
for the past three years had been living the life of a sexless
beast of burden in the Kolyma mines. After all that time
without seeing a woman, the bowl of food received from
my hands had reawakened in him human feelings that had
been on the point of extinction.

"Let's have a bit more, my dear!" he begged a few minutes later, coming back at the other side of the hatch. "What a nice girl you are! Say something nice to me, to bring back old times."

He held out the bowl in his huge hand that had once been strong—the hand of a farmer or a stonemason, with a big black thumbnail.

"Thanks, my pretty one—pray God you'll see your children again one day."

I leaned through the hatch, drew his head toward me, and kissed him on his toothless mouth, with its prickly stubble.

Next morning Verka repeated more often than usual the dread command "Fall out! On the left!" A large party of us politicals was being made up to work in the taiga, and I was one of the first to be detailed. I don't know if this was part of Ahmet's revenge: more probably I was already on the list for Elgen, the notorious "state farm" which we dreaded most of all and to which almost all of us were sent sooner or later.

I had just time to scribble a note for Helmut and hand it to someone who was going to do kitchen work in the men's camp; but I never heard whether he got it, or what became of that chivalrous dishwasher, who forfeited the comfort and safety of his job for my sake. . . .

7

All this time, I was writing cheerful letters to my mother.
My letter smuggled out of the transit camp began: "You
know how much I love traveling, and so I'm pleased that
we are now going on from Vladivostok." Similarly, in the
letters entrusted to my guesthouse friends at Magadan I
gave her fairly articulate descriptions of the northern
scene, ending always by saying that we expected to be
traveling farther. . . . She, poor thing, wrote in reply:
"I keep looking at the map and wondering how there can
be anywhere farther for you to travel to."

I thought of these words as our convoy proceeded along
the road from Magadan to Elgen. We did indeed seem to
have reached the back of beyond, yet we kept on traveling, or rather being transported in open trucks, packed
together and stiff with cold, like sheep to the slaughter.
Nor was there any end to the icy wastes or the sugar-loaf
hills that hemmed us in.

As always happened at the beginning of such a ride, one
or two of us began to make literary comparisons: in this
case Alaska and Jack London's *White Fang*. But very soon
we all fell silent, struck dumb by cold and by the knowledge that the thing we dreaded had come upon us—trans-
portation to Elgen, which had hung over us like a sword
of Damocles throughout our eight months at Magadan.

It was the fourth of April, but there was still a forty-
degree frost and a stiff breeze. The only sign of the approach of spring was the blinding splendor of the pure

snow and the iridescent play of the sun's rays upon it, from which we could not tear our eyes. Alas, we did not yet know that the word "blinding" was literally true: the fairy-tale beauty was treacherous, and the reflection of ultra-violet rays from the snow did indeed make people blind. Acute inflammation of the eyes and conjunctivitis were among the horrors in store for us.

The sense of being at the end of the earth and cut off from civilization made us acutely miserable throughout the journey.

"You know," whispered the woman next to me through chattering teeth, cowering as low as possible against the wind, "it wouldn't surprise me in the least if a mammoth came out from behind that hill." I agreed with her. I too felt that we were getting farther and farther away, not only from towns but from the present era, toward the neolithic age.

Elgen was covered in thick mist as we drove along its main street toward the low wooden building which housed the state farm officials' administration. It was the time of the midday break, and long lines of workers surrounded by guards trudged past us on their way to the camp. The guards' white, quilted sheepskin coats stood out like patches of light against the gray background. The workers, as if at a word of command, turned their heads to look at us, and we too, shaking off the fatigue and stupor of the journey, looked intently at the faces of our new comrades.

"I thought Elgen was only for women, but don't some of these look like men to you?"

"Yes, I think so—but it's hard to tell."

At first we made a joke of the fact that we couldn't tell the difference any longer. But as we got a closer look at the passing ranks, we no longer thought it was funny. They were indeed sexless, these workers in padded breeches and footwear made of cloth, with caps pulled down low over their eyes, and rags covering the lower part of their brick-red, frostbitten faces.

We were appalled by this discovery, and tears burst from many eyes that had long been dry. So that was what we could expect here in Elgen—we who had already lost our professional standing, our rights as Party members and citizens, and our families, were to lose our sex as well. From tomorrow onward we should join the ghostly parade of strange beings who were now tramping through the rock-hard snow toward us.

"Elgen is the Yakut word for 'dead,'" said one of our company—a woman who had been here before, had managed by some miracle to get back to Magadan, and had then been caught associating with a man outside the camp. She pointed out the various buildings of the state farm to us, the stables, the storehouse, and somewhere in the distance the dairy. But these vivid, cheerful-sounding terms seemed to have so little connection with the drab, mournful landscape that we forgot them instantly. What did stick in our minds was the meaning of the native name, and how well it fitted the place. . . .

We came to the women's compound. Barbed wire, symmetrical watchtowers, creaking gates greedily opening their jaws to receive us. Rows of low huts covered with ragged tar board. A single long latrine made of planks, under which rose hummocks of frozen excrement.

Nevertheless, we were glad to have arrived at a habitation of some sort, to see the homely-looking motionless wisp of smoke above the hut in which we were to live. By degrees we lost the sense of intolerable nakedness and defenselessness which had held us in its grip during our journey across the icy, prehistoric land.

Before long we were huddling around the stove on

which a huge caldron of water was boiling. The hut
smelled of drying footcloths and pieces of bread being
toasted. Our new home . . . We slowly undid our rags
and dug into our ration of bread with twisted fingers that
felt as if they were made of glass.

At this dismal time Fate sent us one of those people who
seem to have been put into the world expressly to bring
comfort to those around them. She was the prisoner in
charge of our hut, Marya Dogadkina: a dark-skinned
woman of about fifty, straightforward and quick on her
feet, who spoke an old-fashioned Russian of the Moscow
type. Her method was not to use kind words; on the
contrary, she was always reproving us for something or
other.

"That's a fine way to close the door!" she shouted,
plunging into the thick cloud of icy fog which surrounded
the entrance to the hut, and somehow forcing the warped
and ice-sheeted door into closing properly so that it kept
the warmth in.

"How do you expect your things to dry if they're all
rolled into a ball like that? Didn't your mother teach you
anything?" she said picking up somebody's rags, unrolling
them, and hanging them near the stove on a line on which
one would not have thought there was an inch of space
left.

"Don't take such great mouthfuls of bread, as if you
were a sea gull. If you bolt it down like that you'll be
hungry again in no time. Here, let me toast it for you."
She briskly did so on an improvised iron spit and returned
the warm, fragrant morsel to its owner.

She darted about the hut like quicksilver, dispensing
advice and help in her kindly, fussy, maternal way. We
felt as if we were all guests of hers—the bed and board

might not be luxurious, but what she had, she gave gladly. We thought of her not so much as a fellow prisoner (though she had in fact been convicted of "anti-Soviet agitation") but as a kind of hostess whose task was to make life easier and more endurable for us all.

"I've been expecting you all day, I suppose you were held up by the melting snow. Drink all the hot water you like, it'll warm you. If you haven't got mugs of your own, take them from that shelf. Don't go to the outside latrine at night, you'll get frozen. I've fixed up a big bucket in the corner and I'll empty it in the morning, the guards won't pay attention. And don't get too downhearted—I've been three years here in Elgen, and I'm still alive. It's not as black as it's painted. Have a good sleep now and you'll feel better. It's late. 'The sandman's coming. . . .' "

These last words made me thrill with joy—they were part of a lullaby that our nurse Fima used to sing to little Vaska. As I went to sleep in the upper bunk I felt a strange sensation of peace and homely security. As if in a dream I heard Marya sweeping the floor, clattering with buckets, and miraculously transforming the prison hut into a peasant cottage—a dirty, poverty-stricken one, of course, with black cockroaches scuttling about, but one which none the less smelled of home and fresh-baked bread. As sleep stole over me I heard Fima repeating to my baby son: "The sandman comes with stealthy tread, And all good boys should be in bed" . . .

Next morning, grim reality once more bared its teeth at us. The word "convoy" kept cropping up again. What— was there really a worse place one could be sent to from here? Yes, indeed—there was Mylga, to which one was sent from Elgen as a "punishment," and then there was Izvestkovaya, to which one was sent as a "punishment" from Mylga. And then there were places in the taiga, where people were sent to fell wood, in comparison with which this hut was a palace. And in summer there was haymaking on rough ground, if we lived to see it. . . .

Marya was not one for telling us fairy tales, and it was no use closing one's eyes to reality. But people did live, even in the taiga, and not all the overseers were brutes, some of them were quite decent. . . . Of course our sentences were stiff, worse than the CRTA's even, but things would get better for us in time. At first the higher-ups cracked down hard on the CRTA's, but then one of them had even been given a soft job and put in charge of the bathhouse.

Yes, we politicals were here simply by virtue of our conviction for "terrorism," but for the rest the camp population consisted mostly of hard-boiled multiple offenders —and "mothers." For some reason, Marya told us, the authorities had decided that this was an excellent place in which to build a home for prisoners' children of all ages— who, if they survived, grew up so tough that even a bullet wouldn't kill them. As for the "mothers," this was the term for all female prisoners who had been caught in love affairs and found to be pregnant. They were treated very strictly but not without a certain humanity, if one can call it that. Several times a day the command "Feeding time!" would be heard from the watchtowers, and the muffled sexless figures, guarded by the warders in sheepskin jackets, would stumble in ranks of five toward the children's home. Here each was given her own offspring, which was hard put to it to extract a few drops of milk from the breast of a woman fed on Elgen rations and employed on "land improvement." Usually, after a few weeks, the camp doctor would report that lactation had ceased: the mother would then be sent out to fell timber or harvest the hay, and the child would have to try to survive on what baby food was available. The turnover of these "unmarried mothers" employed as wet nurses was fearful, and their ranks were constantly refilled with transgressors from all over Kolyma.

"Talk about Protection of Mothers and Children!" exclaimed Nina Gviniashvili as she saw for the first time a platoon of mothers surrounded by soldiers with rifles at the ready.

But it was not until later that we learned all these idyllic details about the "Children's Home." At the moment, after our short stay in the hut, we were terribly shaken by rumors filtering out of the administration that timber-felling parties were already being made up. Accounts differed as to which locality was the worst. Some said you could live longer at Kilometer 7 than, say, 14, as the guards were not quite such swine there; others, however, maintained that at 7 you were much more likely to die because it was as cold in the huts there as in the open air.

One of our number, Galya Stadnikova, ventured to ask whether she would be allowed to work at her own specialty as a trained nurse and midwife. The overseer smiled grimly and rapped out: "There are two specialties for people with your sentence: land improvement and tree felling."

I was assigned to Kilometer 7. We were a mixed lot: mostly politicals, but also some of the worst type of criminals and a group of Orthodox Christians, collective farm women from the Voronezh region who had refused to work on Sundays.

We were kept waiting for some time at the camp gates while our guards argued with "Doctor" Kucherenko, a man with a bronzed face who was respectfully addressed as "Doctor" but who, it later turned out, had been a medical orderly in the army and now was in charge of the infirmary. We heard him say in a loud voice: "What if some of them die on the way? Just look how they're dressed." . . . Yes, as usual we politicals were worse off than anyone else. The peasant women had managed to keep their own coarse scarves, and some of the ordinary criminals even had sheepskin coats. We, on the other hand, had not a rag of our own, and our footwear was full of holes which let in the snow.

Later we learned that in the humane-sounding official phrase we were supposed to be "seasonably dressed and shod." If an excessive number of us died, the infirmary staff were apt to get into trouble. This had already happened to Kucherenko, and that was the only reason he was now protesting against our being sent to the taiga.

We stood freezing for more than an hour while the argument went on, accompanied by songs from the ordinary criminals, who hopped around as they bawled at the top of their voices:

We don't work on Saturday,
On Saturday we don't work,
And every day is a Saturday for us.
Ha-ha-ha!

At last, "humanitarianism" triumphed. Kucherenko had evidently succeeded in establishing that we were not seasonably clothed for marching. And so we were taken out there on tractor-drawn trailers, as no other transport could get through to Kilometer 7, which lay off the main track in the depths of the almost virgin taiga.

On we went—across ravines and streams, with the guards cursing and the female criminals uttering their usual obscenities. . . . Kilometer 7, it turned out, was not a very precise definition; it was 7 and a bit, quite a large bit, more. Not a human being, not an animal crossed our path. It was winter still, deepest winter, although the calendar said April—April 1940.

8

• *Tree felling*

Our overseer was a criminal called Kostik, nicknamed the Actor, and a man of some education. At one period of his hectic career he had worked as a stage hand in a provincial theater, and this had added to his vocabulary such words as "mise-en-scène," "farce," and "travesty," which added a distinctive quality to his obscene language.

He regarded us as an absolutely hopeless lot. Surveying our ranks like a commander before a battle, he looked very grim as he cast his eyes over the line of ragamuffins armed with saws and axes. Clearly, if he wanted any fun he would have to go to Elgen for it. He had no use for the criminal types who might have venereal disease, the religious women were obviously cracked, and as for the politicals— well, perhaps they had been women once, but they weren't much to look at now, just walking skeletons—a real travesty, you might say. Pushing back his forelock, he started to sing:

Not a single one for me—
Damn it, what a travesty!

When he saw Pavel Keyzin, the man in charge of timber supplies at the state farm, approaching, he changed his tune and began to talk about the work plan.

"How do you think I can get a day's quota out of this lot, eh?"

Keyzin looked with equal dissatisfaction at us and our rusty, blunted saws. "Who do they think they're sending us?" Though he had not been turned into a sadist by his difficult job, this man had come to regard people as adjuncts to their tools.

It was about two and a half miles from the shacks they had put us in to our place of work. We trudged in single file through the virgin forest, covered deep in snow which became softer during the day. After the first few steps our feet were wet through. When the afternoon frost set in, our footwear became stiff and we could hardly walk for the sharp pain in our frostbitten feet.

Keyzin left us after the first day, and our guards mostly remained smoking around their campfire, so that Kostik had us to himself. His instruction in the art of tree felling was somewhat perfunctory:

"Ever seen a tree like this? No? Christ, what a set of spoiled babes you are. Well, you see the way the snow's piled up around the trunk? The first thing you have to do is to stamp on it till it's firm, like this. . . ."

It was easy enough for him to do so in his strong, smart felt boots, with his breeches tucked into the tops, as the criminals always wore them. When Galya and I tried to follow suit, our footwear once more filled with snow.

"Now make a cut with your ax on the near side, and after that you can start sawing. I suppose you two fine ladies have never seen a saw in your lives. Christ, what a pantomime!"

"Do you really expect Galya and me to fell a tree that size?"

"Not just one," came the curt answer—not from Kostik but from Keyzin, who had come up right behind him. "Eight cubic meters a day's the norm for the two of you."

Kostik, who could not have cared less about either us or the trees, now put in nastily, trying to suck up to his superior:

"Yes, and you have three days to get your hand in. You'll get full rations for that long, but afterward it'll depend on your output. You'll only eat as much as you earn."

For three days Galya and I struggled to do the impossible. Poor trees—how they must have suffered at being mangled by our inexpert hands! Half dead ourselves and completely unskilled, we were in no condition to tackle them. The ax would slip and send a shower of chips into our faces. We sawed feverishly and jerkily, mentally accusing each other of clumsiness—we knew we could not afford the luxury of a quarrel. Time and again the saw got stuck. But the most terrifying moment was when the battered tree was at last on the point of falling, only we didn't know which way. Once Galya got hit on the head, but the medical orderly refused even to put iodine on the cut, saying: "Aha, that's an old trick! Trying to get exempted the first day, are you?"

The religious women from Voronezh, whom we watched closely, seemed to us to possess some kind of magic secret. How quickly and neatly they made the first cuts with the ax! How smoothly and rhythmically they worked the saw! And how obediently the tree fell in the required direction at the feet of these women, used to manual labor from their childhood!

If we had had a chance to get the hang of things and had been properly fed, perhaps we might in time have got to the stage of fulfilling this norm which always seemed just out of our reach. But then the head of our armed guard, who by all accounts was not a complete swine, was switched to Kilometer 14 and was replaced by the head of the guard there. This monster arrived with several of his underlings, who instituted a reign of terror aimed at our destruction.

"This isn't a seaside resort!" he began in terms only too familiar to us. "You've got your norm to fulfill, and you'll be fed according to output. For sabotage you'll go to the punishment cell."

He reminded me a little of Vulturidze in Yaroslavl. There was nothing Caucasian about his pockmarked face and blond eyebrows, but the expression on his face and the way he pursed his lips when he talked to us gave him a family resemblance. We all noticed it.

"He's the living image of Vulturidze! Must be his brother, or at any rate his cousin." So that's what we called him—the Cousin.

Our working day began when we were roused at 5 A.M., our sides aching from the untrimmed logs of which our bunks were made. We also felt a devastating sense of emptiness, which it was necessary to overcome in order to stagger across to the iron stove and fish out our footcloths and mittens from the stinking heap. This was none too easy, because for the first time since the *Dzhurma* we were sharing quarters with common criminals, who thought nothing of stealing other people's footwear, pushing us away from the stove, or grabbing a sharper saw than their own out of someone's hand. It was no use complaining to Cousin: he was exclusively concerned with output, and made it very plain to us at parades and roll calls that no exception would be made for anybody and that he did not intend to throw away precious food on traitors who could not fulfill their norm. In reply to any questions concerned with the conditions we were living in, he put on his Vulturidze expression and snapped back: "This isn't a seaside resort."

And so our life in the forest became one of famine. Maybe Kostik would have taken pity on us and lightened our lot in some way, but Cousin was merciless. He personally saw to it that his underlings kept us away from the campfires and made sure that the overseers did their duty. When Kostik came with his yardstick to measure our daily output, an armed guard stood behind watching him, so that he could not have done anything to help us, even had he wished.

"Eighteen per cent—that's all the takings for today," he would say gloomily, and with a sidelong glance at his escort, put this figure down against mine and Galya's names.

After receiving the morsel of bread corresponding to our "output," we were led out next day to our place of work, literally staggering from weakness. We divided the morsel into two parts, eating one in the morning with boiling water and the other in the forest, sprinkled with snow.

"Don't you think, Galya, that a snow sandwich is much more satisfying than dry bread?"

"Of course it is!"

During the first week on starvation rations we still occasionally made jokes. For instance, as we dragged ourselves home in our filthy rags, bent double and with the skin peeling off our weather-beaten faces, we would make up stories about ourselves from the society page of an imaginary Western journal.

"A gay troop of Amazons returned the other afternoon

from a delightful *fête champêtre* on the grounds of the Château d'Elgen in the Vale of Tuscany." (Elgen was in the Taskan region.) "Shady corners of the woods echoed to their gay voices. By general consent, the most elegant member of the cavalcade was the Russian Princess Zatmilova." (The patches on Galya Zatmilova's breeches indeed looked more bizarre than anybody else's.) "Baroness von Axenburg" (this was a combination of my married and maiden names) "wore a creation inspired by Paquin which will undoubtedly set the trend for the coming season. On returning to the Château, the ladies partook of an exquisite supper of fresh lobsters, under the attentive surveillance of the veteran majordomo, Cousin de Vulturidze. . . ."

For the first few days of our starvation diet, this sort of nonsense helped to keep up our spirits and remind us that we were human. But before long we were in no mood for joking. Cousin brought into effect his second weapon: the non-fulfillment of the day's norm—and practically all the "politicals" were too weak to fulfill it—was treated as sabotage and penalized by confinement in the punishment cell. This was an unheated shack which resembled a public lavatory more than anything else, especially as we were not allowed to go outside to relieve ourselves and no bucket was provided either. At night we had to take turns sitting on the three logs, fastened together, which served as a bunk, so that we had to stand most of the night. We were herded here straight from the forest, wet and hungry, at about eight in the evening and marched straight back into the forest at about five the next morning.

This time, we felt, death had really caught up with us —it was touch and go whether we could survive. We were already quite exhausted trying to escape it. When I saw

myself dimly reflected in a scrap of looking glass which
Galya had found, I quoted to her Marina Tsvetayeva's *
words: "Such a self I cannot live with; such a self I cannot
love." Surely this could not be my own face! Galya did
not waste time trying to comfort me but said with dry
eyes: "I hope he won't abandon the child." She was talking about her husband, who, for some reason, had escaped
arrest.

9

• *Salvation from heaven*

At first we tried to save ourselves—aided by advice from
Kostik the Actor, who was not such a bad fellow after all.

"At the end of your tether, eh?" he said once to Galya
and me at a moment when the guards were not looking.
"Just about ready to give up the ghost, and then curtains,
eh?"

"Well, what else can we do? Just you tell us."

"You've got to use your brains. Listen—only three things
count in Kolyma: swearing, thieving, and window dressing. You just make your choice," he said cryptically.

After this theoretical introduction, we were given an
object lesson by one of our own number, Polina Melni-

---

* Russian poet, 1892–1941, who emigrated in 1922 to France and returned
to Russia in 1939, where she hanged herself.

kova, who by some miracle used to fulfill her norm, working on her own with a one-handed saw. One day we found ourselves working alongside her. Or rather, she was not working but had been sitting for about an hour on a frozen stump: she was huddled up in her rags and had thrown her ax and saw away from her.

"How on earth can she fulfill her norm if she just sits about like that?"

When we asked her, she said that she had fulfilled it already. As we pressed her further, she looked around furtively and then explained:

"This forest is full of piles of timber cut by previous work gangs. No one ever counted how many there are."

"Yes, but anyone can see that they're not freshly cut."

"The only reason you can see is that the cross sections are dark in color. If you saw off a small section at each end, it looks as if it had just been cut. Then you stack them up in another place, and there's your norm. . . ."

This trick, which we christened "freshening up the sandwiches," saved our lives for the time being. We varied Polina's technique slightly: we laid the foundation of our pile with trees we had really cut down ourselves, leaving a couple or so we had felled but not yet sawn up to create the impression that we were hard at it. Then we went to fetch some of the old logs, "freshening up" their ends and stacking them up on our pile. I may add that we did not feel the slightest compunction about resorting to what Kostik called "window dressing." I can't say whether he realized why our output had suddenly increased, but at least he made no comment.

Alas, our respite was short-lived. We had hardly had time to recover from the punishment cell and get used to full rations when the tractors arrived at Kilometer 7 in order to cart off all the timber that was ready. In three days, our "reserves" for the fulfillment of future norms had disappeared. We were now hoist by our own petard. Cousin flew into a rage when he found that our performance had sunk to 18 or 20 per cent of the norm, and again condemned us to the punishment cell for sabotage.

During the eighteen years of our ordeal, many times I found myself face to face with death, but it was an experience I never got used to. Each time, I felt the same frozen horror and made the same frantic efforts to escape. Each time, my indestructibly healthy body found some miraculous way of preserving the flicker of life from extinction. What is more, each time something intervened, something at first sight accidental, but which was really a manifestation of that Supreme Good which, in spite of everything, rules the world. . . .

To begin with, salvation from death in the Elgen forests came to me from cranberries, sour, bitter, northern berries, not ripening at the end of summer as they would do in a normal climate, but remaining from the previous year, to be coaxed out of their hiding place by the timid Kolyma spring, after their ten months' sleep under the snow.

It was already May when, as I was crouching close to the ground in order to cut the branches off a felled larch tree, I noticed in the thawed patch near the stump that miracle of nature—a sprig with five or six berries on it, of a red so deep that they looked almost black, and so tender that it broke one's heart to look at them. As with all over-ripe beauties, they could be destroyed by the slightest touch, however careful. If you tried to pick them, they burst in your fingers; but you could lie on the ground and suck them off the branch with your dried, chapped lips, crushing each one separately against your palate and savoring its flavor. Their taste was indescribable, like that of an old wine—and not to be compared with ordinary cranberries: its sweetness and heady flavor were those of victory over suffering and winter. What a discovery! . . . I ate the first two clusters all by myself; only on finding a third one did I remember my fellow creatures and call excitedly to Galya: "Throw away your ax and come here! I've found 'berries of golden wine.' " With this quotation from Severyanin * I described my treasure trove.

From then on we went into the forest not in despair but in hope. We had discovered that the berries grew mostly on hummocks and around tree stumps, and we managed to find some almost every day. We seriously believed, and eagerly assured each other, that this daily mouthful of "real vitamins" did much to improve our health—we seemed to get less giddy, and the gums around our loosened teeth seemed to bleed less often.

In that mortally dangerous spring, we were also much sustained by the strength of character shown by the semiliterate peasant women from Voronezh who had been sentenced for practicing their religion. Easter that year came at the end of April. Although they always filled their norm without any "window dressing" and it was largely thanks to them that the output plan for Kilometer 7 was fulfilled, Cousin wouldn't even listen to them when they begged to be excused from work for the first day of Easter. They had said that they would make up for it three times over.

"We don't recognize any religious holidays here, and I won't stand for any subversion. Get out into the forest at once, and don't you try any of your tricks! If you do,

* Igor Severyanin, 1887–1941, Futurist poet.

you'll get a punishment you won't forget in a hurry!" And he gave precise instructions to his underlings.

When the women refused to leave their hut, saying repeatedly, "It's Easter, it's Easter, it's a sin to work on Easter Day," they were driven out with rifle butts. When they got to the forest clearing they made a neat pile of their axes and saws, sat down quietly on the frozen tree stumps, and began to sing hymns. Thereupon the guards, evidently on instructions, ordered them to take off their shoes and stand barefoot in the icy water of one of the forest pools, which was still covered with a thin sheet of ice.

I remember that, as we looked on, an old Bolshevik named Masha Mino stood up for the women, shouting furiously at the soldiers:

"What are you doing? These are peasant women—how dare you turn them against the government like that? We are witnesses, and we'll see that you are called to account!"

The soldiers replied with threats and even shots in the air. I do not remember how long this ordeal went on. For the peasant women it was physical torture while we suffered only morally. They went on chanting as they stood in the icy water and we, dropping our tools, ran from one guard to another crying and sobbing, begging and beseeching.

That night the punishment cell was jammed so tight that there was hardly standing room. Nevertheless the hours passed quickly, as we spent almost the whole time arguing about the Voronezh women's behavior: should it be regarded as unenlightened fanaticism or as fortitude in defense of freedom of conscience? Were we to think of them as mad, or should we admire them? And, most crucial of

all: would we have had the courage to do likewise? . . .
In the heat of the argument we almost forgot our hunger
and exhaustion and the damp, stinking cell itself. The most
remarkable fact was that not a single one of the women
got sick from standing in the ice, and next day they fulfilled their norm by 120 per cent. . . .

Some of us pinned our hopes of salvation on the local
medical orderly, whose professional experience—as was not
unusual in the labor camps—was limited to that of a vet-
erinary assistant on a collective farm. The shack in which
he lived, adjoining the log cabin in which the guards were
quartered, was called the "clinic," but the likes of us were
not allowed to enter it: when someone knocked, he came
out onto the porch and stuck a thermometer into the
patient's mouth. His attitude toward "terrorists" was the
same as Cousin's. In order to be exempted from work, one
had to have a temperature of at least forty degrees—he
didn't believe in "spoiling" us either. Moreover, he used
up all the exemptions he was empowered to give for the
benefit of the common criminals, who repaid him either
in food or in kind—for he was still lusty enough despite
his fifty-odd years.

Nevertheless, it was from this quarter that salvation
eventually came to me. To be exact, I was saved by a
surgeon from Leningrad, a prisoner named Vasily Petukhov, who in June 1940 came to Kilometer 7 together with
Kucherenko, the head of the infirmary at Elgen, for the
purpose of conducting a medical inspection. Their appearance was greeted with joy throughout the camp. For some
it might mean a transfer to indoor work, for others a period of "convalescence" at Elgen, where they would not
have to work for two or even three weeks and would be
given increased rations. Even for those who remained in

the forest, there was a prospect of some relief: for inspections did not take place out of the blue, but only when mortality among prisoners was becoming excessive and it was necessary, in the interest of output, to fatten the workers up a bit, like cattle.

This time, I was again lucky. Kucherenko, after feeling my muscles with the air of an expert, went out for some reason and left me alone with Dr. Petukhov. We looked at each other silently for a while. In that sick-bay shack, with its trestle bed covered with cushions and the walls decorated with "artistic" postcards, I beheld the intelligent, cultured face of a real doctor. It was like a message from the civilized world that I thought I had left behind me for ever.

"You're not from Leningrad, are you?" he asked in a low voice.

"No, but my elder son's there now, with relatives."

It turned out that Dr. Petukhov was a good friend of my Leningrad relative, Dr. Fyodorov, who was also a surgeon.

"Wait a minute—a boy aged twelve or thirteen—Alyosha, isn't it? Yes, I saw him with Fyodorov at the beginning of '38, just before I was arrested. I can see, now, your eyes are the same as his."

I laughed and cried—I felt like embracing this stranger who for the moment was closer to me than anyone else in the world, because he had seen Alyosha only two years ago. It was over three years since I had seen my children.

"I'll save you," said the doctor firmly. "Don't think I'm out of my mind, I really mean it. Just think how much better you'd be as a nurse than this medic here. . . . You'd really be able to help people. You know Latin, don't you? Well, that's all you need. That always makes a tremendous impression on Kucherenko. I'll see that you're transferred to the camp."

He made out a three-day exemption from work for me on the ground of "alimentary dystrophy," and I experienced the bliss of lying in an empty hut with a book in my hands—yes, a book, which Lelka, one of the common criminals, had got from the guards. She was fond of me and knew what would give me most pleasure. It was an elementary school textbook—evidently some of our tormentors took correspondence courses.

"The gods," I read over and over again, "lived on Mount Olympus; they drank nectar and ate ambrosia." . . . How glorious it was to lie here instead of sawing wood, and to find that letters formed themselves into words just as they had always done. "They drank nectar"—that must have been like the cranberries buried under the snow. And ambrosia? Oh, that must have tasted like fried potatoes!

Toward evening on my second day of bliss, I heard the rattle of the camp tractor.

"Bring your things!"

A study in dialectic! Those words, which could be so ominous, now had the ring of deliverance: for those of us who heard them knew that we were leaving Kilometer 7, the forest, Cousin and all his works. . . .

"They're making you a medical attendant in the children's home," said the young guard who accompanied us, in a friendly tone. I could have kissed him.

On the way to Elgen, our trailer became unhitched and overturned with us into a ditch. Although it was June, the water was icy. But what did it matter? Once again, I had given death the slip.

All that this book describes is over and done with. I, and thousands like me, have lived to see the Twentieth and the Twenty-second Party Congress.

In 1937, when this tale begins, I was a little over thirty. Now I am in my fifties. The intervening eighteen years were spent "there."

During those years I experienced many conflicting feelings, but the dominant one was that of amazement. Was all this imaginable—was it really happening, could it be intended? Perhaps it was this very amazement which helped to keep me alive. I was not only a victim, but an observer also. What, I kept saying to myself, will come of this? Can such things just happen and be done with, unattended by retribution?

Many a time, my thoughts were taken off my own sufferings by the keen interest which I felt in the unusual aspects of life and of human nature which unfolded around me. I strove to remember all these things in the hope of recounting them to honest people and true Communists, such as I was sure would listen to me one day.

When I wrote this record, I thought of it as a letter to my grandson. I supposed that by 1980, when he would be twenty years old, these matters might seem remote enough to be safely divulged. How wonderful that I was mistaken, and that the great Leninist truths have again come into their own in our country and Party! Today the people can already be told of the things that have been and shall be no more.

Here, then, is the story of an ordinary Communist woman during the period of the "personality cult."

"Alice, do you see that?" I pointed at the screen. "It's the storm," I said. I glanced up again at the boiling clouds. "It's trying to force its way into the starting grid."

"Like it's alive," she said.

Great green spheres tumbled within the gathering cloud, jerking and whirling as if caught in the frenetic grip of madmen on speed. Some of the spheres splintered, revealing blobs of quivering yellow black sludge, like melted wax, or another part of my mind screamed, like putrefying flesh. The sludge ran together and massed towards the starting grid, as if attempting to form some hideous monstrosity, probing at the edges of our sanctuary. Before each new terrible form could congeal, it fell back before beginning anew to grow into something even more appalling, like sentient waves of chaos lapping at the foundations of reality...

Emerald lightning stabbed from the nearest thunderhead, leaping across the threshold, breaking my fascination with the green globes and what they disgorged. The bolt transfixed Sanders, straight through his forehead. Like a fishing line of laser light, the tendril pulled the man upright. Sanders' eyes and mouth snapped open. His irises flamed the same green as the lightning falling all around us....

BY THE SAME AUTHOR

Oath of Nerull
Lady of Poison: The Priests
Darkvision
Stardeep
Plague of Spells
City of Torment
Key of Stars
Sword of the Gods
Spinner of Lies